NMR

Basic Principles and Progress
Grundlagen und Fortschritte

Volume 1

Editors: P. Diehl E. Fluck R. Kosfeld

With 53 Figures

Springer-Verlag Berlin Heidelberg GmbH 1969

Professor Dr. P. DIEHL
Physikalisches Institut der Universität Basel

Professor Dr. E. FLUCK
Institut für Anorganische Chemie der Universität Stuttgart

Dozent Dr. R. KOSFELD
Institut für Physikalische Chemie
der Rhein.-Westf. Technischen Hochschule Aachen

ISBN 978-3-662-12602-8   ISBN 978-3-662-12600-4 (eBook)
DOI 10.1007/978-3-662-12600-4

Title No. 3361

Preface

Nuclear magnetic resonance spectroscopy, which has evolved only within the last 20 years, has become one of the very important tools in chemistry and physics. The literature on its theory and application has grown immensely and a comprehensive and adequate treatment of all branches by one author, or even by several, becomes increasingly difficult.

This series is planned to present articles written by experts working in various fields of nuclear magnetic resonance spectroscopy, and will contain review articles as well as progress reports and original work, its main aim, however, is to fill a gap, existing in literature, by publishing articles written by specialists, which take the reader from the introductory stage to the latest development in the field.

The editors are grateful to the authors for the time and effort spent in writing the articles, and for their invaluable cooperation.

The Editors

Contents

P. Diehl and C. L. Khetrapal

NMR Studies of Molecules Oriented in the Nematic Phase of Liquid
Crystals   1

R. G. Jones

The Use of Symmetry in Nuclear Magnetic Resonance   97

NMR Studies of Molecules Oriented in the Nematic Phase of Liquid Crystals

P. DIEHL and C. L. KHETRAPAL*

Department of Physics, University of Basel, Switzerland

Contents

1. Introduction 3
2. Liquid Crystals 4
    2.1. Classification of Liquid Crystal Phases 4
    2.2. Theories of the Liquid Crystalline State 5
    2.3. Nematic Phases 6
3. Experimental 7
4. Basic Theory (for $I = {}^1/_2$) 7
    4.1. The Hamiltonian of Oriented Spin Systems 8
    4.2. The Direct Coupling (D_{ij}) and the Orientation Matrix $\{S\}$ 8
    4.3. An Alternative Description of Orientation 10
    4.4. The Anisotropy of Chemical Shift 11
    4.5. The Anisotropic Indirect Spin-Spin Coupling Constant (D_{ij}^{ind}) 12
    4.6. The Factors Governing Orientation of Solute Molecules in the Nematic Solvent 12
    4.7. The Influence of Vibrational Motion on the Inter-nuclear Distance ... 14
    4.8. Structural Information Contained in Spectra of Partially Oriented Molecules 15
5. Basic Theory for $I > {}^1/_2$ with Special Reference to $I = 1$ 15
6. Basic Principles of Spectral Analysis 16
    6.1. Nomenclature of the Spectra 16
    6.2. Methods of Analysis 17
    6.3. Limitations in Obtainable Information 18
7. Practical Applications 19
    7.1. Spin Systems which do not Provide Information on the Structure of Molecules 19
        7.1.1. Two Spin Systems (AB, AX and A_2) 19
        7.1.2. Three Spin Systems (ABC, AA′A″) 23

* Permanent address: Tata Institute of Fundamental Research, Bombay (India).

7.1.3. Three Spin System with C_2-Symmetry (AB_2) 26
7.1.4. Three Spin System with C_3-Symmetry (A_3) 30
7.1.5. The System of Four Spins (AB_3) without C_3-Symmetry ... 31
7.1.6. The System of Five Spins (A_3B_2) 33
7.1.7. First Order Spectra of 7 and 10 Spins 36
7.2. Systems Providing Information on the Geometry of Molecules .. 37
7.2.1. Systems of Four Spins $(AB_3$ and $AX_3)$ with C_3-Symmetry . 37
7.2.2. The System of Four Spins $(AA'BB')$ with C_{2v}-Symmetry .. 42
7.2.3. The System $AA'A''A'''$ with D_{2d}-Symmetry 46
7.2.4. The System $AA'XX'$ with C_{2v}-Symmetry 46
7.2.5. The System $AA'A''A'''$ with D_{2h}-Symmetry 47
7.2.6. The System of the Type AB_2C 52
7.2.7. The Five Spin System $(AA'BB'C)$ 54
7.2.8. The System $AA'BB'X$ 56
7.2.9. The Six Spin System $AA'A''A'''A''''A'''''$ with C_6-Symmetry 57
7.2.10. The System of the $A_3A_3'(A_3X_3)$ Type with C_3-Symmetry .. 58
7.2.11. The $AA'A''A'''A''''A'''''$ Type with D_{3h}-Symmetry 62
7.2.12. The System $AA'A''XX'X''$ with D_{3h}-Symmetry 64
7.2.13. The System of the Type $AA'A''A'''XX'$ with D_{2h}-Symmetry 65
7.2.14. The System $AA'BB'CC'$ with C_{2v}-Symmetry 67
7.2.15. The System $AA'BB'CX$ with C_{2v}-Symmetry 69
7.2.16. The System A_3B_2X 72
7.2.17. The 8-Spin System with D_{2d}-Symmetry 72
7.2.18. The $AA'A''A'''BB'B''B'''$ with C_{2v}-Symmetry 74
7.2.19. The 12-Spin System with T_d-Symmetry 74

8. Determination of Absolute Signs of Indirect Coupling Constants 75

9. Spectra of Molecules Dissolved in a Nematic Phase Oriented by an Electric Field 76

10. Anisotropy of Chemical Shift 79
    10.1. Introduction 79
    10.2. Difficulties in the Measurement of Shift Anisotropies 80
    10.3. Results and Discussion 81

11. Anisotropy of the Indirect Coupling Constant 84

12. Information from D and ^{35}Cl Magnetic Resonance Spectroscopy 84
    12.1. Determination of Orientation and Structural Parameters 84
    12.2. Determination of Quadrupole Coupling Constants 85

13. Papers which Appeared after the Manuscript was Submitted 86
    13.1. The System $AA'BB'XX'$ with C_{2v}-Symmetry 86
    13.2. The System $ABB'CXX'$ with C_{2v}-Symmetry 87

Acknowledgements 87
Liquid Crystals Solvents Used in NMR Spectroscopy (Appendix A) 88
Compounds Studied (Appendix B) 90
References 93

1. Introduction

In nuclear magnetic resonance (NMR) experiments with liquids and gases, the local magnetic fields at the sites of the nuclei, responsible for the various transition frequencies, change with the orientation of the molecules relative to the applied external magnetic field. As a consequence of rapid molecular motion only the average values of the shift and coupling parameters can be observed. For nuclei with spin $^1/_2$, each of the chemical shifts and the indirect spin-spin coupling constants, which are actually second rank tensors with as many as nine independent components, affect the normal high resolution NMR spectra (in isotropic media) as single parameters proportional to the trace of the tensors. This simplifies the normal NMR spectra but at the same time imposes a considerable loss of information. Furthermore, in isotropic media, the direct magnetic dipole-dipole interaction transmitted through space becomes zero when averaged. This interaction, proportional to the inverse cube of the inter-nuclear distance, is a symmetric and traceless tensor having a maximum of five independent components. The geometrical information which it carries is also lost as a consequence of the averaging process.

Similarly, the interaction of the nuclear electric quadrupole moment with the local electric field gradient for nuclei with spin $> ^1/_2$ averages zero in the normal high resolution spectra. The direct spin-spin coupling as well as the quadrupole interaction affect the spectra through relaxation processes only.

In solid state NMR both interactions have been directly observed since the earliest experiments, providing data on the crystal structure. However, intermolecular contributions limit the precision in this area. Several years ago, this situation induced NMR spectroscopists to search for possibilities of molecular orientation in which the direct intermolecular dipole-dipole coupling still averages zero while the intramolecular coupling is finite. This was thought to allow the study of direct dipolar and quadrupolar interactions as well as anisotropies of the indirect couplings and chemical shifts. Several types of experiments, e.g. application of an electric field [1], adsorption of molecules in polymers with subsequent stretching, or embedding in zeolite crystal channels [2] were devised. These methods met with only limited success until the use of nematic phases of liquid crystals as orienting solvents was introduced by SAUPE and ENGLERT [3, 4] in 1963.

Since then the spectroscopy of oriented molecules has developed rapidly, the primary objective being the study of direct spin-spin couplings which provide information on relative internuclear distances, bond angles and absolute signs of indirect coupling constants. In the meantime, several reviews on the subject [5 – 10] have appeared but the rapid development warrants the consolidation of the existing data on this fascinating and by no means difficult field of NMR spectroscopy.

2. Liquid Crystals

It was observed as early as 1888 that certain organic substances have two distinct melting points [11]. For example, cholesteryl benzoate turns into a cloudy liquid at 145°C and becomes clear at 179°C. The cloudy intermediate phase was found to have a crystal-like structure and the name 'liquid crystal' was suggested for it. Liquid crystals usually consist of rod-like molecules containing benzene rings. The intermolecular forces tend to orient the molecules in the liquid crystal phase with their longest axes parallel.

2.1. Classification of Liquid Crystal Phases

There are three basic types of liquid crystalline phases: smectic, nematic and cholesteric. They differ in the degree and type of local ordering. Molecules in the smectic phase are 'parallel' and aligned in layers (Fig. 1). Due to the thermal

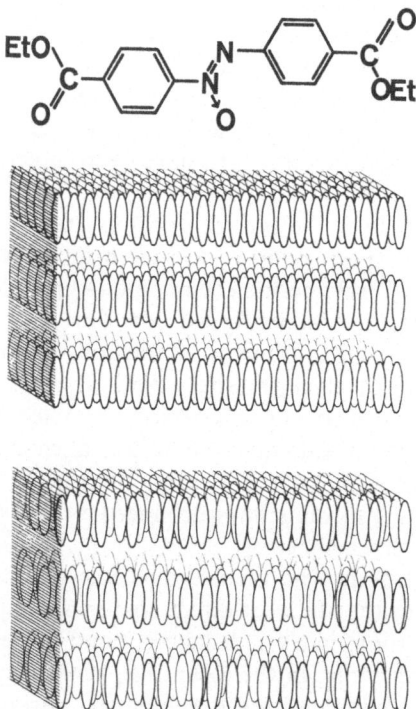

Fig. 1. The structure of a smectic phase, e. g. 4-4′-di-ethyl azoxybenzoate [9]. (Reprinted from Österreichische Chemiker-Zeitung, **4**, 115 (1967), copyright (1967). Reprinted by permission of the copyright owner)

energy of the molecules, however, there is a certain scatter in the orientation of the longest molecular axes. The nematic phase is ordered to a lesser extent than the smectic (Fig. 2) since there is no separation into layers. The cholesteric phase (Fig. 3) resembles the nematic phase. The direction of the longest mole-

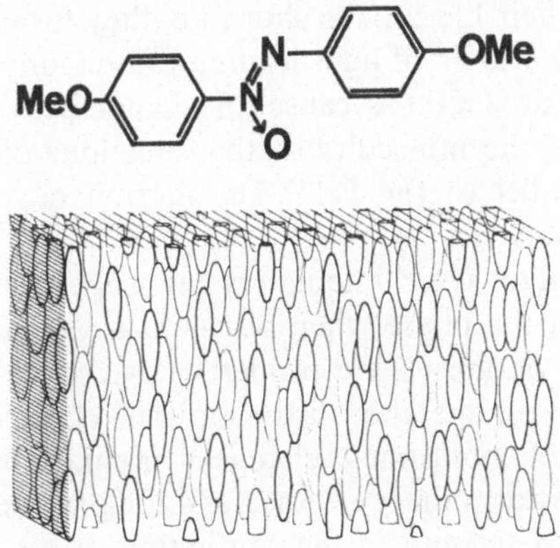

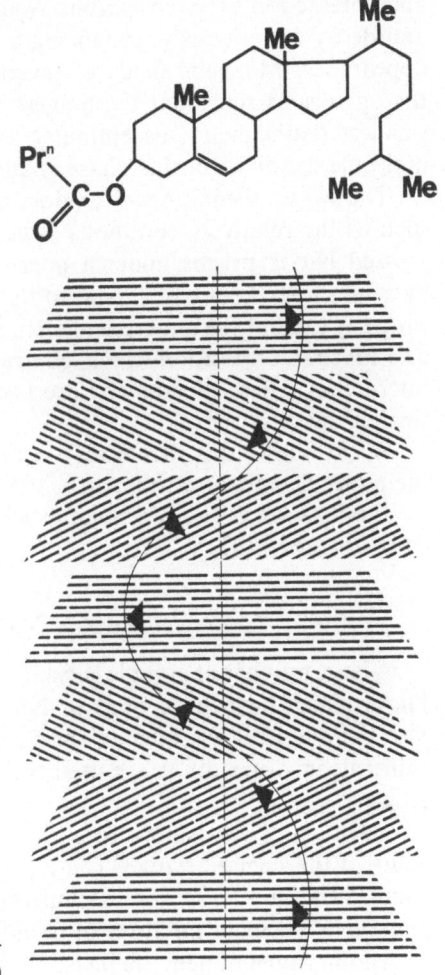

Fig. 2. Arrangement of molecules in a nematic phase, e. g. 4-4'-di-methoxy azoxy benzene [4]. (Reprinted from Österreichische Chemiker-Zeitung, **4**, 115 (1967), copyright (1967). Reprinted by permission of the copyright owner)

Fig. 3. The molecular arrangement in a cholesteric mesophase, e. g. cholesteryl propionate [9]. (Reprinted from Österreichische Chemiker-Zeitung, **4**, 115 (1967), copyright (1967). Reprinted by permission of the copyright owner)

cular axis in each layer is slightly displaced with respect to the neighbouring layers. The overall displacement follows a helical path (Fig. 3). A mixture of two cholesteric compounds of opposite rotatory powers may produce a nematic phase [12–15].

Of the three liquid crystalline phases, the nematic assumes the greatest importance in NMR spectroscopy. All further discussion is devoted to this type.

2.2. Theories of the Liquid Crystalline State

Two main theories have been proposed for a description of the liquid crystalline state; these are the swarm [16] and the distortion or continuum [17] theories.

According to the swarm theory, large numbers of liquid crystal molecules (approximately 10^5) aggregate in swarms. Within these clusters the molecules are arranged 'parallel' primarily due to dispersion forces, but there is no appreciable interaction between various swarms. Their life time is short, i.e. they form and decay continuously, producing a strong scatter of light leading to a cloudy appearance. Magnetic fields of several hundred Gauss cause an alignment of the aggregates such that the longest axes of the molecules or the directions of smallest diamagnetic susceptibility are parallel to the field. The motion of a molecule dissolved in this phase is anisotropic.

The swarm theory, however, does not explain several liquid crystal properties such as the relatively continous structure of the phase when observed between crossed Nicol prisms under a microscope, or the elastic properties evidenced by a return of the liquid crystal to its original shape after distortion. The elastic property can only be understood on the basis of a continuous change of orientation as a function of position which, contrary to swarm theory, indicates a long range interaction. These phenomena are taken into account in the continuum or distortion theory.

According to this theory also, a magnetic field orients the liquid crystal and the molecules dissolved therein.

Equivalence of the two theories when applied to NMR experiments has been demonstrated [18].

2.3. Nematic Phases

A large number of organic substances exhibit liquid crystalline properties [19]. Those which have been used in NMR are compiled in Appendix A together with the temperature range of their nematic state in the pure form. The Roman numerals in Appendix A are used to identify the liquid crystals throughout this article.

Addition of any solute to a liquid crystal usually lowers the temperature limits of the nematic range. The liquid crystal (II) for instance may be used at about $50°C$ and (I) at room temperature. A similar depression of the nematic range may occur in mixtures of liquid crystals [20, 21].

Another type of nematic phase is obtained by mixing cholesteric compounds of opposite optical rotatory powers. A $1 \cdot 9 : 1$ mixture by weight of cholesteryl chloride (*l*-rotatory) and cholesteryl myristate (*d*-rotatory) used as a solvent gives a characteristic spectrum of dissolved benzene (partially oriented) at $40°C$ [15]. The nematic range may be shifted between $23°C$ and $130°C$ by a suitable choice of cholesteryl esters.

A transition from the cholesteric to the nematic phase by application of a magnetic field has been observed [15]. The cholesteric mixture of 2.4 mole percent of optically active amyloxyazoxy benzene in the liquid crystal (III) provides a solvent displaying a typical nematic phase spectrum for dissolved benzene at about $104°C$ [15]. Similarly poly-γ-benzyl-L-glutamate, which forms a cholesteric phase in solutions above 12% of the polymer, has been used for observing the usual spectra of oriented molecules [22–24] such as methylene chloride, dimethylformamide and *p*-xylene in a nematic phase.

A further type of nematic phase, the lyotropic mesophase, formed by a mixture of C_8 or C_{10} alkyl sulphates plus the corresponding alcohols, sodium sulphate and water (or heavy water) in approximate ratio $8:1:1:10$ respectively, has also been suggested [25] and used [26, 27] between $10°C$ and $75°C$.

A liquid crystal molecule generally contains a large number of protons. Due to complex direct dipole-dipole interaction between the many protons, the NMR spectrum of the pure nematic phase usually gives an unresolved spectrum with little observable fine structure. The molecules which are dissolved in the nematic phase, however, give relatively sharp lines and the direct dipole-dipole interaction is manifested by line splitting in the spectra [3].

3. Experimental

NMR spectra of molecules dissolved in the nematic phase of liquid crystals can be studied with the aid of a standard high resolution spectrometer equipped with a variable temperature assembly. Ordinarily a solution of about 20 mole percent concentration of the solute in the liquid crystal is prepared. The mixture is heated above the clearing point in order to obtain a homogeneous solution. The spectrum is then recorded at the appropriate lower temperature.

Line-widths generally range from 4 to 60 Hz. They are often narrower near the centre of the spectrum than in the wings. This may be attributed to the super-imposed effects of temperature-, concentration- and field-gradients. Line-width as well as degree of orientation are temperature and concentration dependent. Reduced temperature and concentration usually produce sharper lines because of smaller gradients [28]. Low concentrations, on the other hand, reduce the signal to noise ratio. The temperature gradients may in practice be lowered by using pure or mixed liquid crystals [20, 21] which form nematic solutions at room temperature.

Since the long molecular axes of the liquid crystalline nematic phase are oriented parallel to the applied magnetic field, rapid rotation of the sample about an axis perpendicular to the field destroys the orientation. Most of the experiments are therefore performed without spinning the sample. This introduces an additional contribution to the line-widths due to field inhomogeneities. In certain cases, it has been observed that slow spinning of the sample is possible without destruction of orientation; the degree of orientation, however, is reduced with increasing speed of rotation [28]. For lyotropic phases as well as for oriented liquid crystals in magnetic fields produced by superconducting magnets or for orientations produced by electric fields applied perpendicular to the magnetic field, the alignment of the mesophase is such that it is possible to spin the sample without destruction of the orientation [26, 29]. Line-widths of the order of $1-5$ Hz have thus been obtained.

4. Basic Theory (for $I = \frac{1}{2}$)

High resolution NMR spectra of molecules dissolved in a nematic phase are generally several kHz wide. If the chemical shifts are negligible, the spectra are symmetrical about the centre.

The spectra are theoretically well understood [1] and the orientation is interpreted in terms of an ordering matrix $\{S\}$ [30]. An alternative notation for the orientation has also been used and the anisotropic motion of the solute molecules expressed as an expansion of a probability function in terms of real spherical harmonics [31]. Both notations are discussed in the following section.

4.1. The Hamiltonian of Oriented Spin Systems

It has already been pointed out that the spectra of oriented molecules depend upon the direct dipole-dipole coupling (D_{ij}), the indirect coupling (J_{ij}) and the chemical shift $(1 - \sigma_i - \sigma_{ia}) \cdot v_0$. The Hamiltonian $(\mathscr{H})$ (eq. 1) of the oriented system differs from that of the isotropic one by additional terms due to direct couplings and the anisotropy of the chemical shift:

$$\mathscr{H} = -\left[\sum_i (1 - \sigma_i - \sigma_{ia}) v_0 \tilde{I}_{zi} + \sum_{i<j}\sum (J_{ij} + 2 D_{ij}) \tilde{I}_{zi} \tilde{I}_{zj} \right.$$
$$\left. + (^1/_2) \sum_{i<j}\sum (J_{ij} - D_{ij})(\tilde{I}_i^+ \tilde{I}_j^- + \tilde{I}_i^- \tilde{I}_j^+)\right]. \tag{1}$$

In eq. (1), J_{ij} and σ_i represent one third of the traces of the corresponding tensors and are identical to the indirect spin-spin coupling and chemical shift observed in normal high resolution NMR spectra in isotropic media. σ_{ia}, the anisotropy of the chemical shift, and D_{ij}, the direct spin-spin coupling, are more complex quantities. They are discussed in the text below.

Eq. (1) shows that the Hamiltonian of an isotropic case transforms easily into that of an oriented system. In the diagonal and off-diagonal contributions to all elements of the matrix, J_{ij} has been replaced by $(J_{ij} + 2 D_{ij})$ and $(J_{ij} - D_{ij})$, respectively, and $(1 - \sigma_i - \sigma_{ia})$ substituted for $(1 - \sigma_i)$.

4.2. The Direct Coupling (D_{ij}) and the Orientation Matrix $\{S\}$

The direct coupling D_{ij} may contain a contribution $D_{ij}^{\text{ind.}}$ from the anisotropy of the indirect spin-spin coupling such that:

$$D_{ij} = D_{ij}^{\text{ind.}} + D_{ij}^{\text{dir.}}. \tag{2}$$

$D_{ij}^{\text{ind.}}$ has the same directional dependence as the direct coupling $(D_{ij}^{\text{dir.}})$. It is therefore called 'pseudo dipolar coupling'. For a pair of protons this type of coupling is small or negligible because of the spherical symmetry of the predominant contact term. For fluorine, on the other hand, the electron-nuclear dipole interactions and the electron orbital current-nuclear dipole contributions to the coupling become significant. These anisotropic interactions give rise to a measurable pseudo dipolar $F-F$ coupling (D_{FF}^{ind}) [32–33]. When the interacting nuclei are protons, D_{ij} is set equal to $D_{ij}^{\text{dir.}}$ if the assumed and observed geometries coincide. Discrepancies caused by intermolecular motion have been observed [34] and are denoted by D_{ij}^{pseudo}.

$D_{ij}^{\text{dir.}}$ is defined by eq. (3):

$$D_{ij}^{\text{dir.}} = -\frac{h\gamma_i\gamma_j}{4\pi^2}\cdot(^1/_2)\left\langle\frac{3\cos^2\theta_{ij}-1}{r_{ij}^3}\right\rangle \tag{3}$$

where θ_{ij} is the angle between the magnetic field direction and the axis connecting the two nuclei i and j separated by a distance r_{ij}, γ is the magnetogyric ratio; and the average is taken over the inter- and intra-molecular motion. If r_{ij} is measured in Angstroms (Å) and D_{ij} in kHz, then the constant $(h\gamma_i\gamma_j/4\pi^2)$ is equal to 120.067 kHz Å³ for a pair of protons. For a proton and a fluorine, it is 112.955 kHz Å³; and for a pair of fluorines, its value is 106.265 kHz Å³.

If nuclei i and j belong to the same rigid part of the molecule and the optic axis of the liquid crystal is parallel to the magnetic field, (as is assumed in this article unless otherwise stated), eq. (3) can be written as:

$$D_{ij}^{\text{dir.}} = -\frac{h\gamma_i\gamma_j}{4\pi^2 r_{ij}^3}\cdot S_{ij} \tag{4}$$

where S_{ij} is the degree of orientation of the axis passing through i and j, and r_{ij} is constant, meaning that the influence of vibrational motions is neglected.

The S-values of different molecular axes are inter-dependent. The average orientation of a rigid molecule is described by a symmetrical and traceless matrix $\{S\}$ with five independent elements [30]. Using ξ, η and ζ as the axes of a Cartesian coordinate system fixed within the molecule and θ_ξ, θ_η and θ_ζ as the angles between these molecular axes and the magnetic field direction, the definition of the matrix elements is given by (5):

$$S_{pq} = (^1/_2)\ \langle 3\cos\theta_p\cos\theta_q - \delta_{pq}\rangle \tag{5}$$
$$p, q = \xi, \eta, \zeta$$

where δ_{pq} is the Kronecker delta ($\delta_{pq} = 1$ for $p = q$, $\delta_{pq} = 0$ for $p \neq q$).

The matrix elements S_{pq} are related to the S-values of an axis, forming the angles α_ξ^a, α_η^a and α_ζ^a with the molecule-fixed coordinate system according to eq. (6):

$$S_a = \sum_{p,q}\cos\alpha_p^a\ \cos\alpha_q^a\cdot S_{pq}. \tag{6}$$

This shows that the degree of orientation of any axis can be derived if $\{S\}$ is known and that the matrix elements S_{pq} may be obtained given sufficient S_a values.

Eq. (5) defines the range of S as $-0.5 \leqslant S \leqslant +1$. At $S_a = 1$ the a-axis is parallel to the applied magnetic field direction, and at $S_a = -0.5$ perpendicular to it. Moreover, the absolute sign of the orientation parameter is positive when the value is larger than 0.5.

By a suitable choice of molecular axes, the number of independent S-values necessary for the description of orientation can be reduced from five to one depending upon the symmetry of the molecule. If the axis is 3-fold or more, its selection as the z-axis causes S_{zz} to be the only independent orientation parameter. If there are two perpendicular planes of symmetry both parallel to the z-axis, a choice of x and y axes in these planes leaves only S_{zz} and $S_{xx}-S_{yy}$. If there is only one plane of symmetry, S_{xx}, S_{yy} and S_{xy} are all independent. It should be noted that in this case, contrary to normal NMR, the plane which contains all interacting nuclei constitutes a plane of symmetry. Table 1 gives the number of independent S-values as a function of molecular symmetry.

Table 1. *Number of parameters necessary for the description of orientation*
$$(S_{xx} + S_{yy} + S_{zz} = 0)$$

Symmetry of the molecule	Number of independent elements of the S-matrix	Independent elements of the S-matrix
3-fold or greater axis	1	S_{zz}
2 perpendicular planes	2	$S_{zz}, S_{xx}-S_{yy}$
1 plane	3	S_{xx}, S_{yy}, S_{xy}
none	5	$S_{xx}, S_{yy}, S_{xy}, S_{xz}, S_{yz}$

The S-matrix obtained experimentally may be diagonalized so that the final number of orientation parameters does not exceed 2. This corresponds to a suitable molecular coordinate system which in general must be assigned empirically.

A more pictorial representation of orientation is obtained by describing the anisotropic motion of the molecules with a probability function $P(\theta, \varphi)$ expanded in real spherical harmonics. $P(\theta, \varphi)$ is assumed to be the probability per unit solid angle that the applied magnetic field direction is θ and φ in spherical polar coordinates relative to the molecule-fixed Cartesian coordinate system [31]. The expansion is given in formula (7):

$$P(\theta, \varphi) = (1/4\pi)\{1 + (\tfrac{5}{2})[(3\cos^2\theta - 1)\cdot S_{zz} + \sin^2\theta \cdot \cos 2\varphi \cdot (S_{xx}-S_{yy})$$
$$+ 4\sin\theta \cdot \cos\theta \cdot \cos\varphi \cdot S_{xz} + 4\sin\theta \cdot \cos\theta \cdot \sin\varphi \cdot S_{yz} \tag{7}$$
$$+ 2\sin^2\theta \cdot \sin 2\varphi \cdot S_{xy}]\}.$$

This approach is similar to a truncated expansion of an orientation probability function based on a Boltzman treatment (section 4.6) in which $\exp\{-(q/kT)\cos^2\theta\}$ is set equal to $\{1-(q/kT)\cos^2\theta\}$. Consequently, $P(\theta, \varphi)$ does not rigorously describe the orientation probability except in the case of small orientations. For large orientation parameters, $P(\theta, \varphi)$ may even assume negative values. On the other hand such a truncation cannot be avoided since the NMR-experiments do not provide more information.

4.3. An Alternative Description of Orientation

In an alternative description of the orientation [31], the probability function $P(\theta, \varphi)$ (eq. (7)) is expressed as follows:

$$P(\theta, \varphi) = (1/4\pi) + c_{3z^2-r^2}\cdot D_{3z^2-r^2} + c_{x^2-y^2}\cdot D_{x^2-y^2}$$
$$+ c_{xz}\cdot D_{xz} + c_{yz}\cdot D_{yz} + c_{xy}\cdot D_{xy}. \tag{8}$$

Here,
$$D_{3z^2-r^2} = (\sqrt{5/8}\,\pi)(3\cos^2\theta - 1)$$
$$D_{x^2-y^2} = (\sqrt{15/8}\,\pi)(\sin^2\theta \cdot \cos 2\varphi)$$
$$D_{xz} = (\sqrt{15/4}\,\pi)(\sin\theta \cdot \cos\theta \cdot \cos\varphi) \qquad (9)$$
$$D_{yz} = (\sqrt{15/4}\,\pi)(\sin\theta \cdot \cos\theta \cdot \sin\varphi)$$
$$D_{xy} = (\sqrt{15/8}\,\pi)(\sin^2\theta \cdot \sin 2\varphi)$$

The coefficients $c_{3z^2-r^2}$, $c_{x^2-y^2}$, c_{xy}, c_{yz} and c_{zx} are called motional constants. To these, D_{ij}^{dir} is related by eq. (10):

$$D_{ij}^{\text{dir.}} = -\frac{h\gamma_i\gamma_j}{4\pi^2\sqrt{5}} \cdot \left\{ + c_{3z^2-r^2}\left[\left\langle\frac{(\Delta z_{ij})^2}{r_{ij}^5}\right\rangle_{\text{Av}} - (^1/_2)\left\langle\frac{(\Delta x_{ij})^2}{r_{ij}^5}\right\rangle_{\text{Av}} - (^1/_2)\left\langle\frac{(\Delta y_{ij})^2}{r_{ij}^5}\right\rangle_{\text{Av}}\right]\right.$$
$$+ c_{x^2-y^2}\sqrt{3}\left[(^1/_2)\left\langle\frac{(\Delta x_{ij})^2}{r_{ij}^5}\right\rangle_{\text{Av}} - (^1/_2)\left\langle\frac{(\Delta y_{ij})^2}{r_{ij}^5}\right\rangle_{\text{Av}}\right] + c_{xz}\sqrt{3}\left[\left\langle\frac{(\Delta x_{ij})(\Delta z_{ij})}{r_{ij}^5}\right\rangle_{\text{Av}}\right]$$
$$\left. + c_{yz}\sqrt{3}\left[\left\langle\frac{(\Delta y_{ij})(\Delta z_{ij})}{r_{ij}^5}\right\rangle_{\text{Av}}\right] + c_{xy}\sqrt{3}\left[\left\langle\frac{(\Delta x_{ij})(\Delta y_{ij})}{r_{ij}^5}\right\rangle_{\text{Av}}\right]\right\}. \qquad (10)$$

where $\Delta z_{ij} = z_i - z_j$, $\Delta x_{ij} = x_i - x_j$ and $\Delta y_{ij} = y_i - y_j$ such that x_p, y_p and z_p are the coordinates of the p'th nucleus in the molecule-fixed system and the average is taken over the internal motion. It should be noted that eq. (10) differs from that of reference [31] by a factor of 0.5. This introduces equivalence between the definitions of D_{ij}^{dir} given in references [30] and [31]. Throughout this article values of D_{ij}^{dir} obtained according to [31] have been divided by 2. The relations between the S-values and the motional constants are as follows:

$$c_{3z^2-r^2} = \sqrt{5} \cdot S_{zz} \cdot (^1/_2)(3\cos^2\alpha - 1),$$
$$c_{x^2-y^2} = \sqrt{(\tfrac{5}{3})} \cdot (S_{xx} - S_{yy})(^1/_2)(3\cos^2\alpha - 1),$$
$$c_{xz} = 2\sqrt{(\tfrac{5}{3})} \cdot S_{xz} \cdot (^1/_2)(3\cos^2\alpha - 1), \qquad (11)$$
$$c_{yz} = 2\sqrt{(\tfrac{5}{3})} \cdot S_{yz} \cdot (^1/_2)(3\cos^2\alpha - 1),$$
$$c_{xy} = 2\sqrt{(\tfrac{5}{3})} \cdot S_{xy} \cdot (^1/_2)(3\cos^2\alpha - 1),$$

where α is the angle between the liquid crystal optic axis and the direction of the magnetic field.

4.4. The Anisotropy of Chemical Shift

Analogous to formulae (7) and (8), the anisotropy of chemical shift may be expressed in terms of the shielding tensor components [31]:

$$(\sigma_i + \sigma_{ia}) = (\tfrac{1}{3})(\sigma_{xxi} + \sigma_{yyi} + \sigma_{zzi}) + c_{3z^2-r^2} \cdot (2/3\sqrt{5})[\sigma_{zzi} + (^1/_2)(\sigma_{xxi} + \sigma_{yyi})]$$

$$+ c_{x^2-y^2}\left(\frac{1}{\sqrt{15}}\right)(\sigma_{xxi} - \sigma_{yyi}) + c_{xz}\left(\frac{1}{\sqrt{15}}\right)(\sigma_{xzi} + \sigma_{zxi}) \qquad (12)$$

$$+ c_{yz}\left(\frac{1}{\sqrt{15}}\right)(\sigma_{yzi} + \sigma_{zyi}) + c_{xy} \cdot \left(\frac{1}{\sqrt{15}}\right)(\sigma_{xyi} + \sigma_{yxi}).$$

Substitution of relations (11) in eq. (12) results in eq. (13) for $\alpha = 0$:

$$
\begin{aligned}
(\sigma_i + \sigma_{ia}) = (\tfrac{1}{3})(\sigma_{xxi} + \sigma_{yyi} + \sigma_{zzi}) + (\tfrac{2}{3})[(S_{zz}\,\sigma_{zzi} + S_{yy}\,\sigma_{yyi} \\
+ S_{xx}\,\sigma_{xxi}) + S_{xz} \cdot (\sigma_{xzi} + \sigma_{zxi}) + S_{yz} \cdot (\sigma_{yzi} + \sigma_{zyi}) \\
+ S_{xy}(\sigma_{xyi} + \sigma_{yxi})] .
\end{aligned} \tag{13}
$$

Molecular symmetry reduces the number of non-zero elements of the chemical shift tensor (Table 2). For example if the axis is 3-fold or higher, eq. (13) reduces to (14):

$$
(\sigma_i + \sigma_{ia}) = (\tfrac{1}{3})(\sigma_{xxi} + \sigma_{yyi} + \sigma_{zzi}) + (\tfrac{2}{3})S_{zz} \cdot [\sigma_{zzi} - (^1/_2)(\sigma_{xxi} + \sigma_{yyi})] .
$$

or,
$$
\sigma_{ia} = (\tfrac{2}{3})S_{zz}\,\Delta\sigma_i \tag{14}
$$

where
$$
\Delta\sigma_i = [\sigma_{zzi} - (^1/_2)(\sigma_{xxi} + \sigma_{yyi})] .
$$

Table 2. *Molecular symmetry and the non-zero elements of the chemical shift tensor*

Symmetry of the molecule	Number of non-zero elements of chemical shift tensor	Non-zero elements of chemical shift tensor
3-fold or greater axis	2	$\sigma_{zz}, \sigma_{xx} + \sigma_{yy}$
2 perpendicular planes	3	$\sigma_{zz}, \sigma_{xx} + \sigma_{yy}, \sigma_{xx} - \sigma_{yy}$
1 plane	5	$\sigma_{zz}, \sigma_{xx}, \sigma_{yz}, \sigma_{xz}, \sigma_{zx}$
none	9	$\sigma_{zz}, \sigma_{xx}, \sigma_{yy}, \sigma_{xy}, \sigma_{yx}, \sigma_{xz}, \sigma_{zx}, \sigma_{yz}, \sigma_{zy}.$

4.5. The Anisotropic Indirect Spin-Spin Coupling Constant (D_{ij}^{ind})

By substitution of J for each σ in eqs. (12), (13) and (14), the corresponding relations for the anisotropic indirect coupling constants are obtained. $J_i = \tfrac{1}{3}(J_{xx} + J_{yy} + J_{zz})$ is the isotropic value of the coupling constant, i. e. the indirect coupling. J_{ia} corresponds to $2\,D_{ij}^{\text{ind}}$.

4.6. The Factors Governing Orientation of Solute Molecules in the Nematic Solvent

In principle, dispersion forces, permanent electric dipole moments, steric effects and specific interactions may affect the orientation of the solute molecules in the nematic phase. Of these only the influences of dispersion forces and permanent electric dipole moments have been studied in detail [35−36]. Investigations on a variety of chloro [35] and fluoro [36] benzenes in several solvents at various temperatures and concentrations, have shown that only the former is of major importance. The discussion of the problem has been simplified by assuming apolar orientation, validity of the dipole-dipole approximation and absence of specific solute-solvent interaction [142]. The orientation dependent part (F) of the molecular free energy is then given by eq. (15):

$$
F = -a_1 \cos^2\theta_1 - a_2 \cos^2\theta_2 \tag{15}
$$

where θ is the angle between a molecule fixed coordinate system and the optic axis of the liquid crystal. The orientation parameters may be expressed according to Boltzmann statistics (eq. (16)):

$$S_{ii} = \left(\tfrac{3}{2}\right) \cdot \frac{\int\limits_0^{\pi/2} \int\limits_0^{\pi/2} \cos^2\theta_i \exp\left(-\dfrac{F}{kT}\right) \cdot \sin\theta_1 \cdot d\varphi\, d\theta_1}{\int\limits_0^{\pi/2} \int\limits_0^{\pi/2} \exp\left(-\dfrac{F}{kT}\right) \cdot \sin\theta_1 \, d\varphi\, d\theta_1} - \left(\tfrac{1}{2}\right) \qquad (16)$$

where $i = 1$ or 2 and $\cos\theta_2 = \sin\theta_1 \sin\varphi$.

The physical meaning of the coefficients a_i is readily apparent. They represent the differences in free energy between the two solute orientations for which axes i or axis 3 (perpendicular to i) are parallel to the liquid crystal axis. In order to simplify the treatment further, the following additional assumptions are made:
(A) The coefficients a_i are regarded as products of two factors, one depending upon the properties of the liquid crystal and the other on those of the solute.
(B) The coefficients a_i are sums of additive contributions of separate parts of the solute, e.g. of the σ-bonds (C$-$H or C$-$F).

The validity of these assumptions implies that (a_1/a_2) is independent of the nematic solvent, temperature and concentration.

If it is further assumed that the C$-$H and the C$-$F bonds interact with the liquid crystal as described by formula (17):

$$F \sim - a_\sigma \cos^2 \Omega \qquad (17)$$

where F is the free energy contribution and Ω the angle between the bond axis and the liquid crystal axis, the energy coefficients for chloro and fluoro benzenes are then expressed as follows:

$$a_i = a_B + b \sum \cos^2 \alpha_i \qquad (18)$$

where

$a_B = a$ (for unsubstituted benzene)
α_i = angle between the bond and axis i, and
$b = [a_\sigma(\text{CF}) - a_\sigma(\text{CH})]$.

Eq. (18) provides relations between the energy parameters of the various substituted benzenes. For example, the molecules 1,2- and 1,3-difluorobenzenes should have equal a-values,

$$a_1 = a_B + \tfrac{1}{2} b$$
$$a_2 = a_B + \tfrac{3}{2} b,$$

if the twofold axes of symmetry are taken to be axis 2 in the ortho and axis 1 in the meta compound.

The experiments performed to verify the theory confirm its usefulness [35, 36]. The electric dipole moment contribution is indeed not important.

2*

The parameter (a_1/a_2) varies little with temperature (less than 10% variation per $20°\,C$ change of temperature), with concentration (less than 10% for a change of concentration by a factor of 2) or with nematic solvent (less than 30% difference between two liquid crystals). The a-values for the various substituted benzenes as predicted by eq. (18) agree with experimental values within 30%.

The results on fluoromethanes also [137] indicate that the geometrical factors are more important than the dipole interactions in determining molecular orientation. If J_{HF} is assumed positive, the $C-F$ bond in fluoromethane, the $F-C-F$ plane in methylenefluoride and the F, F, F plane in fluoroform orient preferentially in the direction of the magnetic field.

4.7. The Influence of Vibrational Motion on the Inter-nuclear Distance

The inter-nuclear distance is usually assumed to have the equilibrium value, as pointed out in the discussion of eq. (4). The influence of vibrational motions is neglected. Actually, the measured direct coupling constants are proportional to the vibrational average value $\langle 1/r^3 \rangle$ so that the inter-nuclear distances affecting the spectra are $(\langle r^{-3} \rangle)^{-1/3}$. Since different types of mean values are obtained from various methods of measurements, such as electron diffraction or microwave spectroscopy, the geometry of a molecule derived from NMR in oriented systems may deviate from that of other methods. Whereas the deviations are complex for polyatomic molecules, they may be derived quantitatively for diatomic molecules [37].

Table 3. *Bond lengths of diatomic molecules obtained by different methods* [37]. *The relative deviation from the equilibrium value is:*

$$\xi = (r - r_e)/r_e.$$

For harmonic vibration $\langle \xi \rangle$ is zero.

Technique	Bond lengths	r/r_e	Practical example $(C^{12}-H)$
Electron diffraction	$\langle r \rangle$	$1 + \langle \xi \rangle$	1.1388 Å
Microwave	effective $(\langle r^{-2} \rangle)^{-1/2}$	$1 + \langle \xi \rangle - (^3/_2)\langle \xi^2 \rangle$	1.1307 Å
NMR	$(\langle r^{-3} \rangle)^{-1/3}$	$1 + \langle \xi \rangle - 2\langle \xi^2 \rangle$	1.1280 Å [a] calculated
Microwave	substitution	$1 + f(\langle \xi \rangle - (^3/_2)(\langle \xi^2 \rangle))$	[b] 1.1198 Å (equilibrium value)

[a] From electron diffraction and microwave values.

[b] $f = (\mu/m_1)[1 + (\mu_1/\mu)^{\frac{1}{2}}]^{-1} + (\mu/m_2)(1 + \mu_2/\mu)^{-1}$ where m_1 and m_2 are the masses of the atoms in the parent molecule with reduced mass μ and μ_1 and μ_2 the reduced masses of the isotopically substituted molecules.

In microwave spectroscopy, two types of molecular structures are discussed. They are the r_0-structure, corresponding to a rigid rotor analysis and to effective moments of inertia, and the r_s-structure obtained by the substitution method in which differences between effective moments of inertia of isotopically substituted species are taken into account. The distances r_0 and r_s may deviate from the equilibrium r_e or the average $\langle r \rangle$. Moreover, it should be pointed out that the microwave data are calculated from molecules in the vibrational ground state, whereas the NMR values may also be influenced by low-lying excited vibrational states.

In electron diffraction, one observes $\langle r \rangle$.

The various bond lenghts are compared in Table 3.

Table 3 suggests that the largest differences should be observed between the values obtained by electron diffraction and NMR and the smallest between microwave effective structure and NMR values.

Experiments in various phases such as gas, liquid crystal or solid may also result in disagreement between the observed distances.

4.8. Structural Information Contained in Spectra of Partially Oriented Molecules

Eqs. (4), (7), (8) and (10) show that the spectrum is not affected by a scaling process changing the inter-nuclear distances by a constant factor and modifying the orientation parameters accordingly. NMR spectroscopy of oriented systems consequently does not allow the measurement of absolute distances but does permit the determination of distance ratios. Hence only the shape of the skeleton of nuclei contributing to the spectrum can be studied. Individual orientation parameters can be measured only if either a known inter-nuclear distance or a moment of inertia [64] is introduced.

5. Basic Theory for I > $\frac{1}{2}$ with Special Reference to I = 1

For nuclei with $I > \frac{1}{2}$ a contribution describing the interaction of the electric quadrupole moment of nuclei with the neighbouring electrons and nuclei must be added to the Hamiltonian (eq. 1):

$$\tilde{\mathscr{H}}_Q = \frac{eQ_p}{h} \sum_p \sum_{ij} V_{ij}^p S_{ij} \left[3\tilde{I}_{zp}^2 - I_p(I_p + 1) \right] \cdot \left[4 I_p(2I_p - 1) \right]^{-1}, \tag{19}$$

where (eQ_p) is the electric quadrupole moment of nucleus p

V_{ij}^p = second derivative of the electrostatic potential at p.

I_p = spin of p.

For nuclei with $I = 1$. e. g. deuterons, eq. (19) reduces to (20) [7]:

$$\tilde{\mathscr{H}}_Q = h \sum_p D_{pp} \tilde{I}_{zp}^2 \tag{20}$$

where $D_{pp} = (1/2\,h)\,eQ_p \sum_{ij} V_{ij}^p S_{ij}$ and constant contributions to the Hamiltonian are neglected.

If it is further assumed that the tensor V_{ij} has rotational symmetry and that the z-axis lies in the direction of the bond to the deuteron, D_{pp} is given by eq. (21):

$$D_{pp} = (3/4\,h)\,eQ_p\,V_{zz}^p \cdot S_{zz} \qquad (21)$$

From this it can be seen that the energy levels of deuterons are equally affected for $I_z(D) = \pm 1$ but are unaffected for $I_z(D) = 0$. Consequently the transition energies corresponding to $I_z = (-1 \leftarrow 0)$ and $I_z = (0 \leftarrow +1)$ differ by $2\,D_{pp}$.

The spectrum of an isolated deuteron is, therefore, a doublet with splitting Δ which with eq. (22) yields the quadrupole coupling constant $(eQ\,V_{zz}/h)$ if the orientation parameter is known [38].

$$(eQ\,V_{zz}/h) = (2\,\Delta/3\,S_{zz}) \qquad (22)$$

Measurement of the orientation parameter from proton resonance in the same sample at the same temperature provides the necessary information. It should be noted that the precision of such a quadrupole coupling constant-measurement is related directly to the accuracy of an inter-nuclear distance which is required for the description of orientation.

If, on the other hand, the quadrupole coupling constant is known, the deuteron spectrum allows the determination of an orientation parameter. For large molecules this is much easier from the simple deuteron spectra than from the corresponding complex proton spectra. Due to the small magnetogyric ratio the direct and indirect couplings with deuterons are smaller than with protons.

6. Basic Principles of Spectral Analysis

In the analysis of a spectrum, the parameters are obtained from the observed line positions and intensities. It has been shown that such a determination from the NMR spectra of oriented molecules is unique [39]. Hence, in principle, these parameters may be used to derive unique structural information.

6.1. Nomenclature of the Spectra

The nomenclature used for normal high resolution NMR spectroscopy [40] may be applied to spectra of oriented systems. There are, however, two differences which should be kept in mind. Weak coupling in spectra of isotropic solutions requires that (J_{ij}^2/δ_{ij}) be of the order of the line-width. For oriented systems D_{ij} replaces J_{ij} in this condition. For homonuclei it is rarely fulfilled so that weak coupling is generally confined to heteronuclei. The second important difference is found in the definition of 'magnetic equivalence' [41–43]. In normal NMR, a group which is magnetically equivalent leads to a singlet resonance if (1) the nuclei have the same chemical shift and (2) they are equally coupled to all the nuclei within groups having different chemical shifts. If furthermore, all the nuclei within a group are equally coupled to one another, the group is called 'fully equivalent'. In NMR spectroscopy of oriented molecules such a group with n nuclei leads to a multiplet of lines with spacing equal to $|3\,D|$ and a binomial distribution of intensity.

The absence of full equivalence for benzene in the oriented case requires that the notation A_6 of normal spectroscopy be changed to $AA'A''A'''A''''A'''''$. A further representation uses A_n for full equivalence and numerals within the subscript followed by a multiplication sign for the number of fully equivalent groups therein [41]. The symbol $A_{3 \times 2}B_2$ denotes the proton spectrum of n-propane in anisotropic media. The corresponding notation is A_6B_2 for the spectroscopy in isotropic media.

6.2. Methods of Analysis

The analysis of spectra of the type AB, AB_2 and AB_3 can be carried out 'by hand' since the line positions and intensities are explicit analytical functions of the parameters [1, 44, 45]. Group theory has been applied to obtain analytical expressions for systems with high symmetry such as benzene [3, 4]. The sub-spectral method [46] of analysis has been suggested [47] and applied to total spin [41] and effective Larmor-frequency subspectra [48, 49]. The direct and moment methods have been employed in cases where the line positions alone do not provide sufficient relations for the analysis [50]. A method of reconstruction of the separate components (chemical shift, indirect and direct couplings) of the Hamiltonian from line positions and intensities has been developed [39] and applied to an AB-type spectrum [51]. Spin tickling [51] and double irradiation [52] experiments have also been performed,

The systems $AA'BB'$ [53] and $AA'A''$ [50] can be solved analytically if only the direct couplings between the interacting nuclei are considered.

For more complex systems, in which the line positions cannot be expressed analytically in terms of the parameters, computer methods involving simulation [31 – 33, 54] or iterative procedures [27, 41, 55] have been developed.

The simulation method involves the computation of line positions and intensities assuming a Lorentzian line-shape. The input parameters are the chemical shifts, the indirect spin-spin couplings, the line-width for the assumed shape, the anisotropic motional constants, the nuclear coordinates in Ångstroms and the pseudo dipolar coupling constants. The method allows computation of D_{ij}^{dir} and of the theoretical spectrum. The experimental and simulated spectra are compared and the values of the input parameters varied until agreement between the two spectra is achieved. Since the programme does not obtain the parameters iteratively, a comparison of spectra 'by eye' is necessary which may not produce the best possible fit.

The iterative programmes commonly used are adaptations of the standard LAOCOON II and LAOCN 3 [56]. They take into account the direct as well as indirect couplings together with the chemical shifts [27, 41, 55]. Details on the modification of LAOCOON II have been reported [55] and the programme has been renamed LAOCOONOR. The input parameters are chemical shifts and direct and indirect couplings. It is possible to iterate on all or any combination of the parameters.

A set of D-values is normally calculated from the approximately known geometry and the spectrum is computed with the help of Laocoonor (first part). In the spectrum thus calculated, it is possible to identify some of the transitions with those observed experimentally. On this basis the spectrum is then iteratively computed. This allows the identification of many more lines. Retaining the well resolved transitions, the final iterative computation is performed. The method is quite satisfactory even for relatively complex spectra such as those of pyridine [55] and benzene-d$_1$ [38].

In practice, the indirect coupling constants are usually not iterated upon but are taken from data on NMR spectroscopy in isotropic media.

The existence of the pseudo dipolar coupling may be indicated if the ratios of the direct couplings deviate from the theoretical values defined by molecular geometry. The magnitude, however, can be obtained if the inter-nuclear distances are precisely known. The difference between the theoretical and experimental values of D_{ij}^{dir} is designated D_{ij}^{ind}.

6.3. Limitations in Obtainable Information

It has been mentioned earlier that the spectra of oriented molecules may provide information about the orientation of the solute molecules, the relative inter-nuclear distances, anisotropies of chemical shifts and indirect coupling constants as well as the signs of indirect couplings. The amount of obtainable information, however, depends critically upon molecular symmetry and the number of independent direct couplings. The number of relations which can be derived for molecular structure is equal to the difference between the number of independent direct couplings and the parameters necessary to describe molecular orientation. These relations may or may not be sufficient to define the entire shape of the skeleton of nuclei under study. Thus, a system of less than four spins without symmetry does not provide geometric information. On the other hand, a four spin system with a threefold axis of symmetry [57] requires only one parameter [31] for the specification of orientation but has two direct couplings. Hence, the ratio of the inter-proton distance which completely defines the shape can be derived from the spectrum. A four-spin system without symmetry has only one direct coupling in excess of the number of parameters necessary to define the orientation. The analysis of its spectrum provides one geometric relation which cannot describe the whole shape. The spectra of planar 4, 5 and 6 spin systems with C_{2v}-symmetry can in principle provide the ratios of all the inter-proton distances [53, 55, 38] even though unforeseen difficulties may arise in practice [58, 59]. The equations relating the inter-nuclear distances to the direct couplings in some cases may not provide a single definite solution [58, 59]. Coincidentally the values of the direct couplings may be such that the results of numerical solutions are inaccurate.

A further limitation arises in systems of 4 or less spins due to the fact that the information contained in the transition frequencies is insufficient for the determination of all the parameters. For example, no unique solution may be obtained for AB, ABC and ABCD spectra from line positions alone if indirect couplings are also unknown [50]. The spectrum of a three spin system of the type AA'A" without appreciable indirect couplings has an infinite number of solutions [50] even if line intensities are included.

Another limitation to obtainable information is introduced when the spectra are 'deceptively simple'. In such cases, at least one of the sub-spectra is degenerate. There may be various conditions causing the degeneracy but the most important, arising when $\delta \rightarrow 0$, have been discussed in detail [60]. In the analysis of such spectra, the main difficulty lies in the assignment of lines, the number of which is smaller than that expected theoretically. Lines with nearly similar energy may overlap, intensities of transitions may be too small to be detected, or near perfect symmetry of a spectrum may simulate zero chemical shift.

Conditions for the appearance of deceptively simple spectra of the type ABX, AB_2X, AA'BB'X and AA'XX' have been discussed [60].

7. Practical Applications

The present chapter is divided into two sections, the first dealing with spectra which do not provide information about the geometry of molecules and the second describing the analysis and geometric information obtained from other systems investigated.

7.1. Spin Systems which do not Provide Information on the Structure of Molecules

7.1.1. Two Spin Systems (AB, AX and A_2)

The spectrum of an oriented two spin system of the type AB consists in general of four lines. Expressions for the frequencies and intensities of the four transitions are summarised in Table 4 [1].

Table 4. *Transition frequencies and intensities for an oriented 2-spin system (AB). Frequencies are relative to* $(1/2)(v_{AB} + v_B) = 0$

No.	Transition	Frequency	Relative intensity
1	$1\,s_0 \leftarrow s_1$	$(1/2)(J_{AB} + 2\,D_{AB})$ $+ (1/2)[(J_{AB} - D_{AB})^2 + \delta_{AB}^2]^{1/2}$	$1 - \{(J_{AB} - D_{AB})$ $\cdot [(J_{AB} - D_{AB})^2 + \delta_{AB}^2]^{-1/2}\}$
2	$2\,s_0 \leftarrow s_1$	$(1/2)(J_{AB} + 2\,D_{AB})$ $- (1/2) \cdot [(J_{AB} - D_{AB})^2 + \delta_{AB}^2]^{1/2}$	$1 + \{(J_{AB} - D_{AB})$ $\cdot [(J_{AB} - D_{AB})^2 + \delta_{AB}^2]^{-1/2}\}$
3	$s_{-1} \leftarrow 2\,s_0$	$-(1/2)(J_{AB} + 2\,D_{AB})$ $+ (1/2)[(J_{AB} - D_{AB})^2 + \delta_{AB}^2]^{1/2}$	$1 + \{(J_{AB} - D_{AB})$ $\cdot [(J_{AB} - D_{AB})^2 + \delta_{AB}^2]^{-1/2}\}$
4	$s_{-1} \leftarrow 1\,s_0$	$-(1/2)(J_{AB} + 2\,D_{AB})$ $- (1/2)[(J_{AB} - D_{AB})^2 + \delta_{AB}^2]^{1/2}$	$1 - \{(J_{AB} - D_{AB})$ $[(J_{AB} - D_{AB})^2 + \delta_{AB}^2]^{-1/2}\}$

Table 4 shows that there are only two observable, independent spacings but three unknown parameters (δ_{AB}, J_{AB} and D_{AB}) to be determined from the spectrum. No unique solution of the problem can be derived from line positions alone. Line-intensity must be included in the analysis. This can be done either in terms of the second moment ⌊50⌋ or directly as given in Table 4 and discussed in the following paragraph.

If x and y represent the distances of lines 1 and 2 from the centre and T, the ratio of intensities of the two lines:

$$\left(T = \frac{[(J_{AB} - D_{AB})^2 + \delta_{AB}^2]^{\frac{1}{2}} + (J_{AB} - D_{AB})}{[(J_{AB} - D_{AB})^2 + \delta_{AB}^2]^{\frac{1}{2}} - (J_{AB} - D_{AB})} \right),$$

the parameters δ_{AB}, D_{AB} and J_{AB} are related to these measured quantities as follows:

$$\delta_{AB} = \frac{2(x-y)\sqrt{T}}{(T+1)} \tag{23}$$

$$D_{AB} = \frac{2}{3} \cdot \frac{(Ty + x)}{(T+1)} \tag{24}$$

$$J_{AB} = (x + y) - \frac{4}{3} \cdot \frac{(Ty + x)}{(T+1)}. \tag{25}$$

It can be seen from eqs. (23), (24) and (25) that the solution is not affected by an exchange of x with y. For a fixed x (for example line 1), y can be the distance to the centre from either line 2 or 3. Consequently, there are two solutions to the problem [50]. The correct solution can be inferred from tickling which allows the identification of energy levels.

In practice, J_{AB} values may already be known from NMR in isotropic media. The problem is then simplified and the remaining two parameters can be obtained from line positions only. Under these circumstances, the following two equations are used:

$$D_{AB} = \tfrac{1}{2}(x + y - J_{AB}) \tag{26}$$

$$\delta_{AB}^2 = [(x-y)^2 - (J_{AB} - D_{AB})^2] \tag{27}$$

The spectrum shown in Fig. 4 is of the AB type [51].
The tickling experiment (Fig. 4 B) demonstrates that the repeated spacing should be counted between lines 1 and 2 (Table 4).

Due to a coupling with neighbouring nuclei which are quadrupole relaxed, the proton lines may be broadened selectively leading to an independent determination of the unique solution. For example in Fig. 4, the lines 2 and 4 which are broader than 1 and 3 should be attributed to the same nucleus.

The values of the parameters obtained for 4,6-dichloropyrimidine are listed in Table 5.

The analysis allows the determination of the sign of D_{AB} relative to that of J_{AB}. Because there is only one direct coupling constant, it is not possible to

Table 5. *Available information on the NMR spectra of two spin systems in the nematic phase*

Compound	Structure	Spectral type	Nematic solvent	Concentration (mole %)	Temperature (°C)	Results	Reference
4,6-dichloropyrimidine		AB	1:1 mixture of (I) + (II)	21	27	$D_{AB} = \pm(15.9\pm0.1)$ Hz $J_{AB} = \pm(0.8\pm0.3)$ Hz $\delta_{AB} = (94.1\pm2)$ Hz	[51]
acetylene	HC≡CH	A_2	(III)	—	75	$S_{HH} = -0.014$ $\Delta\sigma = (8.1\pm2.2)$ ppm.	[45]
1,2,3,5-tetrachloro-benzene		A_2	(IV)	—	82	$|S_{HH}| = 0.056$	[4]
1,2,4,5-tetrachloro-benzene		A_2	(III)	—	93	$|S_{HH}| = 0.022$	[4]
2,3,5,6-tetrachloro-fluorobenzene		AX	(IV)	—	77	$|S_{HH}| = 0.052$	[4]

Table 5 (continued)

Compound	Structure	Spectral type	Nematic solvent	Concentration (mole %)	Temperature (°C)	Results	Reference
(cis)1,2-dichlorethylene		A_2	(II)	at various temperatures and concentrations		effects of temperature, concentration and spinning speed on orientation parameters are discussed	[28]
(trans)1,2-dichloro-ethylene		A_2	(II)	"	"	"	[28]
1,1-dichloroethylene		A_2	(II)	"	"	"	[28]
2,4,5-trichloronitro-benzene		AB but effectively A_2 (degenerate)	1:1 mixture of (I) + (II)	17	27	degenerate spectrum, chemical shift lies between 0 and 20 Hz	[60]
methylene chloride	CH_2Cl_2	A_2	poly-γ-ben-zyl-L-gluta-mate	more than 12% polymer	room tempe-rature	doublet observed	[23]

obtain the sign of the orientation parameter, even if it is assumed that the molecule orients preferentially with its plane along the direction of the magnetic field as do other aromatics. Consequently, the absolute sign of J_{AB} cannot be derived.

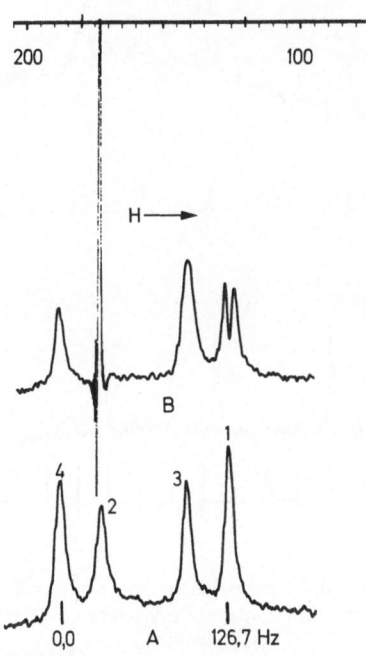

Fig. 4. Proton spectrum of 4,6-dichloro pyrimidine in the nematic phase of a mixture of 50% (I) and 50% (II) [51]. Solute concentration = 21 mole percent. Temperature = 27°C. Spectrometer frequency 60 MHz
(A) Un-irradiated   (B) Tickling (transition 2 irradiated)

Several spectra of the type A_2 and AX have also been studied, and the results are summarised in Table 5.

7.1.2. Three Spin Systems (ABC, AA'A'')

Spectra of the type ABC have no unique solution from line positions if all the parameters are unkonwn. In practice indirect couplings are usually known from the spectra in isotropic solvents. If the chemical shifts and the indirect couplings are negligibly small, there is no solution to the problem [50] even if line intensities are included.

As an example of the ABC class, the spectrum of racemic 3,3,3-trichloropropylene oxide [61] is discussed. The study has been made in a mixture of cholesteryl chloride and cholesteryl myristate and also in the nematic phase of liquid crystal (III) (Fig. 5).

Due to their different orientations, the *d*- and *l*-enantiomers give distinct NMR spectra in the optically active liquid crystal. Doublets of lines in Fig. 5 A confirm this difference. In the nematic phase of (III), only one spectrum is observed (Fig. 5 B).

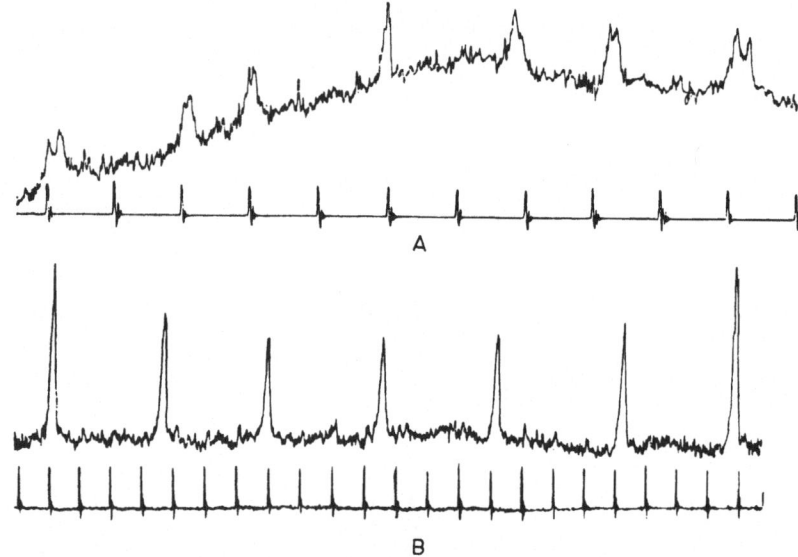

Fig. 5. NMR spectra of 3,3,3-trichloropropylene oxide [*61*] in solvents: (A), 1.9 : 1 (by weight) mixture of cholesteryl chloride and cholesteryl myristate at 40° C. (B), the nematic phase of (III) at 90° C
Spectrometer frequency 60 MHz. Frequency markers shown below the spectra are spaced at 106 Hz
(Reprinted from J. Am. Chem. Soc. **90**, 2183 (1968), copyright (1968) by the American Chemical Society. Reprinted by permission of the copyright owner)

Further, the chemical shifts in the isotropic phase have been found to lie between those in the mixture of cholesteryl esters and those in the liquid crystal (III). This is consistent with the observation that the solvent mixture orients the benzene plane at right angles to the magnetic field if phase (III) orients it along the direction of the applied field [*15*]. The analysis has been performed by the computer simulation method and the values of the parameters obtained are reproduced in Table 6. The indirect couplings used are: $J_{AB} = 4.6$ Hz, $J_{AC} = 3.6$ Hz, $J_{BC} = 2.1$ Hz.

Spectra of oriented systems are usually dominated by the direct couplings, for protons. Three non-equivalent protons without symmetry normally give spectra of the type AA'A''. The transition frequencies and intensities of such systems can be expressed analytically in terms of the parameters (Table 7) [*50*]. Table 7 shows that the analysis of such a spectrum provides the values of A^2 and $(B^2 + C^2)$, from which only $D_{AB} + D_{BC} + D_{AC}$ and $D_{AB}D_{BC} + D_{BC}D_{AC} + D_{AC}D_{AB}$ can be derived. As a result the number of solutions is infinite, even if line intensities are considered.

Table 6. *Parameters obtained from the spectra of racemic 3,3,3-trichloropropylene oxide and 1,2,4-trichlorobenzene*

Compound	Structure	Spectral type	Nematic phase	Parameters obtained (Hz unless otherwise mentioned)	References
racemic 3,3,3-trichloro-propylene oxide		ABC	cholesteryl chloride and cholesteryl myristate mixture	$\delta_B = 0.8$, $\delta_C = -27.5$, $\delta_A = 0.0$ $D_{AC} = 173.75$, $D_{BC} = -7.0$, $D_{AB} = -252.5$ for one enantiomer and $D_{AB} = -260.25$, $D_{AC} = 181.50$ $D_{BC} = -7.75$ for the other	[61]
racemic 3,3,3-trichloro-propylene oxide		ABC	(III)	$\delta_B = -16.0$, $\delta_C = -57.0$ $\delta_A = 0.0$ $D_{AB} = 609.25$, $D_{AC} = -353.5$ $D_{BC} = 13.75$	[61]
1,2,4-trichloro-benzene		ABC but treated as AA'A"	(III)	$D_{AB} + D_{BC} + D_{AC} = 1752$ $D_{AB}D_{BC} + D_{BC}D_{AC} + D_{AC}D_{AB} = 31248$ Hz^2 (J and δ assumed zero)	[50]

Table 7. *Transition frequencies and intensities for an oriented three-spin system without symmetry (AA'A''). Only direct couplings are considered*

Line number	Transition	Frequency	Intensity
1	$s_{3/2} - 1s_{1/2}$	$-(3/2)A + x$	$3x_1^2/(x_1^2 + B^2 + C^2)$
2	$s_{3/2} - 2s_{1/2}$	$-(3/2)A - x$	$3x_2^2/(x_2^2 + B^2 + C^2)$
3	$3s_{1/2} - 3s_{-1/2}$	0	1
4	$1s_{1/2} - 1s_{-1/2}$	0	$\{[2x_2^2 - (B^2 + C^2)]/$ $[x_2^2 + B^2 + C^2]\}^2$
5	$2s_{1/2} - 2s_{-1/2}$	0	$\{[2x_1^2 - (B^2 + C^2)]/$ $[x_1^2 + B^2 + C^2]\}^2$
6	$1s_{1/2} - 2s_{-1/2}$	$-2x$	$[2x_1x_2 - (B^2 + C^2)]^2/$ $[(x_1^2 + B^2 + C^2)(x_2^2 + B^2 + C^2)]$
7	$2s_{1/2} - 1s_{-1/2}$	$+2x$	$[2x_1x_2 - (B^2 + C^2)]^2/$ $[(x_1^2 + B^2 + C^2)(x_2^2 + B^2 + C^2)]$
8	$1s_{-1/2} - s_{-3/2}$	$(3/2)A - x$	$3x_1^2/(x_1^2 + B^2 + C^2)$
9	$2s_{-1/2} - s_{-3/2}$	$(3/2)A + x$	$3x_2^2/(x_2^2 + B^2 + C^2)$

$A = (-1/2)(D_{AB} + D_{BC} + D_{AC}); \quad x = (1/2)(A^2 + 4(B^2 + C^2)]^{1/2}.$
$B = (1/2\sqrt{2})(D_{AB} - 2D_{BC} + D_{AC}); \quad x_1 = (1/2)(A - x).$
$C = (3/2\sqrt{6})(D_{AB} - D_{AC}); \quad x_2 = (1/2)(A + x).$

In this category, the proton spectrum of a 17 mole per cent solution of 1,2,4-trichlorobenzene in the nematic phase of (III) has been studied at 75°C [50]. The general appearence of the spectrum resembles that of Fig. 5B but is more symmetrical. The small asymmetry is due to a slight chemical shift. The quantities determined are included in Table 6.

7.1.3. Three Spin System with C_2-Symmetry (AB$_2$)

An AB$_2$ system has two direct coupling constants. Two parameters are required to describe the molecular orientation.

The standardised Hamiltonian is given in Table 8.

Table 8. *Hamiltonian for oriented three spin system with C_2-symmetry. Standardisation $(v_A + 2v_B) = 0$, $D = D_{AB}$; $D' = D_{BB'}$. In H_{44} and H_{55}, the sign of δ is reversed with respect to H_{22} and H_{33} respectively*

Element	Value
H_{11}	0
H_{22}	$-\frac{2}{3}\delta - J - 2D$
H_{33}	$\frac{1}{3}\delta - (1/2)J - D - (3/2)D'$
H_{23}	$(1/\sqrt{2})(J - D)$
H_{77}	0
H_{88}	$-\frac{2}{3}\delta$

The analysis of the spectrum can be carried out by the 'direct method' [62, 50] or by the normal method [44]. Both are illustrated in the following paragraphs.

For the 'direct analysis' the traces (T) and the determinants (P) are obtained as follows:

$$
\begin{aligned}
T_1 &= \mathrm{E}_{22} + \mathrm{E}_{33} = -\tfrac{1}{3}\delta - \tfrac{3}{2} \cdot J - 3D - \tfrac{3}{2}D', \\
T_2 &= \mathrm{E}_{44} + \mathrm{E}_{55} = \tfrac{1}{3}\delta - \tfrac{3}{2}J - 3D - \tfrac{3}{2}D', \\
P_1 &= \mathrm{E}_{22}\mathrm{E}_{33} = -\tfrac{2}{9}\delta^2 + \delta D' + \tfrac{3}{2}D^2 + 3DJ + 3D'D + \tfrac{3}{2}D'J, \\
P_2 &= \mathrm{E}_{44}\mathrm{E}_{55} = -\tfrac{2}{9}\delta^2 - \delta D' + \tfrac{3}{2}D^2 + 3DJ + 3D'D + \tfrac{3}{2}D'J.
\end{aligned}
\tag{28}
$$

The unknown parameters are given by:

$$
\begin{aligned}
\delta &= \tfrac{3}{2}(T_2 - T_1), \\
D' &= (P_2 - P_1)/2\,\delta, \\
D &= \tfrac{1}{9}\left[K \pm \{K^2 + 9KD' - 4\delta^2 - 9(P_1 + P_2)\}^{1/2}\right], \\
J &= K/3 - 2D, \\
K &= T_1 + T_2 - (P_1 - P_2)/(T_1 - T_2).
\end{aligned}
\tag{29}
$$

Relations (28) and (29) show that the parameters of the system can be determined from known line positions. There are, however, two possible solutions for the value of D. A unique solution can be derived by inclusion of intensity information as the second moment ($\langle \omega^2 \rangle$):

$$
\langle \omega^2 \rangle = \tfrac{2}{3}\delta^2 + \tfrac{3}{2} \cdot D'^2 + 3D^2.
\tag{30}
$$

In the normal method parameters are extracted from expressions for the line positions and intensities reproduced in Table 9 [44].

Table 9. *Frequencies and intensities for a three spin-system of the type* AB_2

Line	Transition	Frequency	Intensity
1	$1s_{1/2} - s_{3/2}$	$(3/4)D_{BB} + (3/2)D_{AB} - (3/4)J_{AB} - \delta/2 + W_1$	$(\sqrt{2}\,a_1 + a_2)^2$
2	$2s_{1/2} - s_{3/2}$	$(3/4)D_{BB} + (3/2)D_{AB} - (3/4)J_{AB} - \delta/2 - W_1$	$(a_1 - \sqrt{2}\,a_2)^2$
3	$1s_{-1/2} - 1s_{1/2}$	$W_2 - W_1$	$(a_1 b_1 + \sqrt{2}\,a_2 b_1 + \sqrt{2}\,a_1 b_2)^2$
4	$1s_{-1/2} - 2s_{1/2}$	$W_2 + W_1$	$(-a_2 b_1 + \sqrt{2}\,a_1 b_1 - \sqrt{2}\,a_2 b_2)^2$
5	$2s_{-1/2} - 1s_{1/2}$	$-W_2 - W_1$	$(-a_1 b_2 + \sqrt{2}\,a_1 b_1 - \sqrt{2}\,a_2 b_2)^2$
6	$2s_{-1/2} - 2s_{1/2}$	$-W_2 + W_1$	$(a_2 b_2 - \sqrt{2}\,a_1 b_2 - \sqrt{2}\,a_2 b_1)^2$
7	$s_{-3/2} - 1s_{-1/2}$	$-(3/4)D_{BB} - (3/2)D_{AB} + (3/4)J_{AB} - \delta/2 - W_2$	$(\sqrt{2}\,b_1 + b_2)^2$
8	$s_{-3/2} - 2s_{-1/2}$	$-(3/4)D_{BB} - (3/2)D_{AB} + (3/4)J_{AB} - \delta/2 + W_2$	$(b_1 - \sqrt{2}\,b_2)^2$
9	$a_{-1/2} - a_{1/2}$	$-\delta$	1

where:

$$
\delta = -v_A \quad \text{and} \quad v_B = 0.
$$

$$
W_1 = \tfrac{1}{2}\left[(\delta)^2 - 2(\delta)D_{AB} + 3(\delta)D_{BB} + (\delta)J_{AB} + \tfrac{3}{2}D_{BB}J_{AB} + \tfrac{9}{4}D_{BB}^2 - 3D_{AB}D_{BB} + 3D_{AB}^2 + 3D_{AB} \cdot J_{AB} + \tfrac{9}{4}J_{AB}^2\right]^{\frac{1}{4}}.
$$

$$W_2 = \tfrac{1}{2}\ [(\delta)^2 + 2\,(\delta)\,D_{AB} - 3\,(\delta)\,D_{BB} - (\delta)\,J_{AB} + \tfrac{3}{2}\,D_{BB}J_{AB} + \tfrac{9}{4}\,D_{BB}^2 - 3\,D_{AB}D_{BB} + 3\,D_{AB}^2 + 3\,D_{AB}J_{AB} + \tfrac{9}{4}J_{AB}^2]^{\frac{1}{2}}$$

$$\frac{a_1}{a_2} = \frac{-\sqrt{2}\,(J_{AB} + D_{AB})}{(\delta + \tfrac{3}{2}D_{BB} - D_{AB} + \tfrac{1}{2}J_{AB} - 2\,W_1)}\ ;\ a_1^2 + a_2^2 = 1$$

$$\frac{b_1}{b_2} = \frac{-\sqrt{2}\,(J_{AB} + D_{AB})}{(-\delta + \tfrac{3}{2}D_{BB} - D_{AB} + \tfrac{1}{2}J_{AB} - 2\,W_2)}\ ;\ b_1^2 + b_2^2 = 1$$

As demonstrated in Table 9 the position of line (9) determines the value δ. Lines (1) and (2) give W_1 and $(D_{BB} + 2D_{AB} - J_{AB})$ whereas (7) and (8) provide $(D_{BB} + 2D_{AB} + J_{AB})$ and W_2. All parameters can be obtained from a combination of these expressions. As discussed earlier, there are two solutions to the problem, the correct one can be deduced by inclusion of intensity data.

The molecules of this type which have been studied are listed in Table 10. A typical spectrum of 3-chloropropyne is shown in Fig. 6.

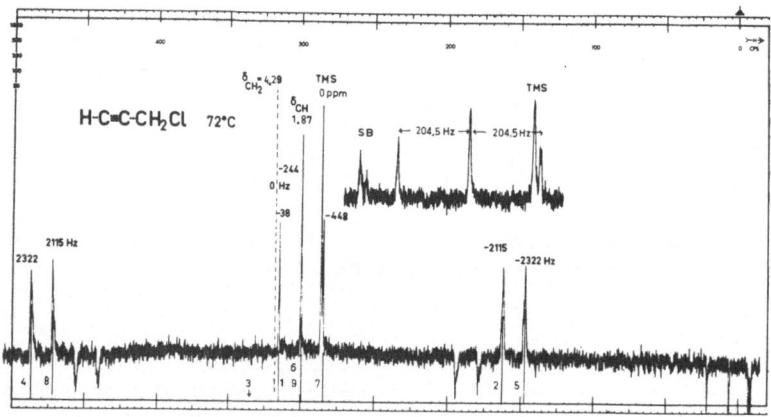

Fig. 6. Proton magnetic resonance spectrum of 3-chloropropyne in the nematic phase of (III) [57]. Concentration = 15 mole %, temperature = 72° C
Spectrometer frequency = 100 MHz. Lines with inverted intensity are n (2001) Hz side bands introduced by the integrator. (Reprinted from Mol. Cryst. 1, 503 (1966), copyright (1966). Reprinted by permission of the copyright owner and the authors)

The orientation of the molecule 1,2,3-trichlorobenzene has been studied in various liquid crystals at several temperatures and concentrations in order to investigate the intermolecular forces responsible for orientation [35] (see section 4.6).

In AA_2' systems there are three parameters $(J_{AA'}, D_{AA'}$ and $D_{A'A'})$ to be derived from the two independent relations of eq. (28). Inclusion of the intensity information in the form of eq. (30) allows the determination of all the parameters. Of the four possible solutions to the fourth order equation two are always imaginary [50].

Table 10. *Results from the $AB_2(AA'_2)$ type of spectra*

Compound	Structure	Spectral type	Nematic phase	Conc. in mole %	Temp. in °C	Results	Reference
3-chloropropyne	$HC{\equiv}C-CH_2Cl$	AB_2	(III)	15	72	$D_{AB} = -100.0\ \text{Hz}$, $D_{BB} = -1476.0\ \text{Hz}$ $J_{AB} = 2.7\ \text{Hz}$, $\nu_B - \nu_A = 244.0\ \text{Hz}$ $S_{C_3} = 0.064$, $S_{CH_2} = -0.061$ $\Delta\sigma_H = \sigma_\parallel - \sigma_\perp = 12.4 \times 10^{-6}$	[57] [45]
3,5-dichloro benzoic acid		AB_2	(V)	15	127	$D_{AB} = 316.0\ \text{Hz}$, $D_{BB} = -541.0\ \text{Hz}$ $\nu_B - \nu_A = -226.0\ \text{Hz}$ $J_{AB} = -1.6\ \text{Hz}$ $S_{11} = 0.55$; $S_{22} = -0.21$	[44]
3-bromopropyne	$HC{\equiv}C-CH_2Br$	AB_2	(III)	–	70	$D_{AB} = -79.5\ \text{Hz}$ $D_{BB} = 1169.0\ \text{Hz}$ $S_{C_3} = 0.051$, $S_{CH_2} = -0.049$ $\Delta\sigma_H = \sigma_\parallel - \sigma_\perp = 12.4 \times 10^{-6}$	[57] [45]
1,2,3-trichloro benzene		AB_2	(III)	6	96	$S_{11} = 0.049$ $S_{22} = 0.133$	[35]
2,6-dibromo pyridine		$\approx A(A'_2)$	(III)	20	80	$D_{BB} = \pm 165\ \text{Hz}$ $D_{AB} = \pm 695\ \text{Hz}$ $J_{AB} = 7\ \text{Hz (assumed)}$ $S_{11} = \mp 0.024$, $S_{22} = \mp 0.109$	[50]

S_{11} is along the C_2-axis and S_{22} perpendicular to it.

3*

Of the remaining two, the correct solution may sometimes be deduced from the molecular geometry together with the limiting values of the orientation parameters.

The spectrum (Fig. 7) of 2,6-dibromopyridine approaches the AA'_2 type [50].

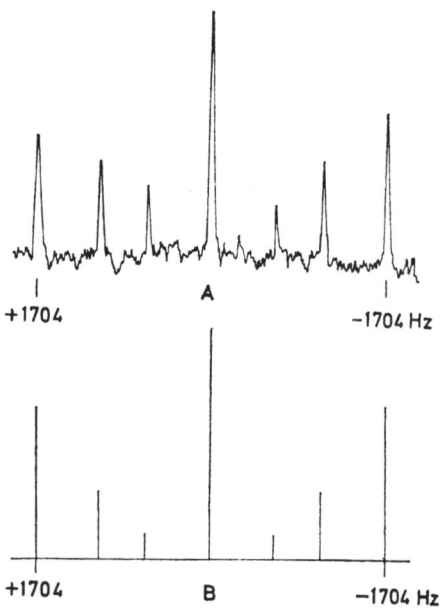

Fig. 7. PMR spectra of 2,6-dibromopyridine in the nematic phase of (III) [50]. A) observed B) calculated
Concentration = 20 mole%, temperature = 80 °C. Spectrometer frequency = 60 MHz

The spectral symmetry indicates that δ is negligible within the experimental error and does not exceed the line-width [60]. The two solutions are:

$$J_{AB} = 7 \text{ Hz (assumed)}$$
$$(1) \quad D_{BB} = \pm 165 \text{ Hz}; \quad D_{AB} = \pm 695 \text{ Hz}$$
$$(2) \quad D_{BB} = \pm 342 \text{ Hz}; \quad D_{AB} = \pm 871 \text{ Hz}.$$

The latter is not permissible since it gives $S > 1$. The correct S-values are included in Table 10.

7.1.4. Three Spin System with C_3-Symmetry (A_3)

A single parameter describes the molecular orientation in a three spin system of the type A_3. The spectrum consists of a triplet with intensity ratios $1:2:1$ and a repeated spacing of $|3\,D|$. Fig. 8 shows a typical spectrum of 1,3,5-trichlorobenzene [4] with $|S_{11}| = |S_{22}| = 0.076$.

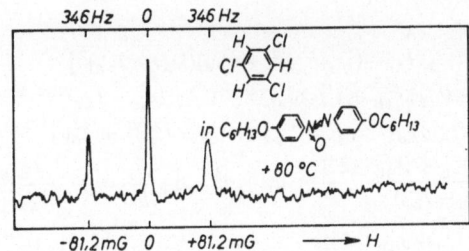

Fig. 8. PMR spectrum of 1,3,5-trichlorobenzene in the nematic phase of (III) [*4*] Temperature = 80°C. Spectrometer frequency = 60 MHz. (Reprinted from Z. Naturforsch. **19a**, 172 (1964), copyright (1964). Reproduced by permission of the copyright owner)

Several other compounds leading to A_3-type spectra or sub-spectra are discussed in section 7.2.1.

7.1.5. The System of Four Spins (AB₃) without C₃-Symmetry

In spectral analysis the treatment of AB_3 type molecules with or without C_3-symmetry axes is identical. However, the latter does not provide structural details.

The expressions for transition frequencies and intensities of an AB_3-spectrum are given in Table 11 [*45*].

Table 11. *Frequencies and intensities of the various transitions for oriented 4-spin systems of the AB_3 type*

No.	Transition	Frequency relative to $v_B = 0$	Intensity
1	$1A_1-A_2$	$-(^3/_2)D_{BB}-2D_{AB}-J_{AB}+\Delta_1-(^1/_2)\delta_{AB}$	$(a_1+\sqrt{3}a_2)^2$
2	$2A_1-A_2$	$-(^3/_2)D_{BB}-2D_{AB}-J_{AB}-\Delta_1-(^1/_2)\delta_{AB}$	$(\sqrt{3}a_1-a_2)^2$
3	$1A_0-1A_1$	$-(^3/_2)D_{BB}+E-\Delta_1$	$(2a_2b_1+\sqrt{3}a_1b_2+a_2b_2)^2$
4	$1A_0-2A_1$	$-(^3/_2)D_{BB}+E+\Delta_1$	$[-\sqrt{3}a_2b_2+a_1(2b_1+b_2)]^2$
5	$2A_0-1A_1$	$-(^3/_2)D_{BB}-E-\Delta_1$	$[\sqrt{3}a_1b_1+a_2(b_1-2b_2)]^2$
6	$2A_0-2A_1$	$-(^3/_2)D_{BB}-E+\Delta_1$	$[-\sqrt{3}a_2b_1+a_1(b_1-2b_2)]^2$
7	$1A_{-1}-2A_0$	$(^3/_2)D_{BB}+E+\Delta_2$	$[-b_2(c_1+\sqrt{3}c_2)+2b_1c_1]^2$
8	$2A_{-1}-2A_0$	$(^3/_2)D_{BB}+E-\Delta_2$	$[b_2(c_2-\sqrt{3}c_1)-2b_1c_2]^2$
9	$1A_{-1}-1A_0$	$(^3/_2)D_{BB}-E+\Delta_2$	$[b_1(c_1+\sqrt{3}c_2)+2b_2c_1]^2$
10	$2A_{-1}-1A_0$	$(^3/_2)D_{BB}-E-\Delta_2$	$[b_1(\sqrt{3}c_1-c_2)-2b_2c_2]^2$
11	$A_{-2}-1A_{-1}$	$(^3/_2)D_{BB}+2D_{AB}+J_{AB}-\Delta_2-(^1/_2)\delta_{AB}$	$(\sqrt{3}c_1+c_2)^2$
12	$A_{-2}-2A_{-1}$	$(^3/_2)D_{BB}+2D_{AB}+J_{AB}+\Delta_2-(^1/_2)\delta_{AB}$	$(c_1-\sqrt{3}c_2)^2$
13	$1E_0-1E_1$	$-D_{AB}-(^1/_2)J_{AB}+F-(^1/_2)\delta_{AB}$	$2(d_1-d_2)^2$
14	$2E_0-E_1$	$-D_{AB}-(^1/_2)J_{AB}-F-(^1/_2)\delta_{AB}$	$2(d_1+d_2)^2$
15	$E_{-1}-1E_0$	$D_{AB}+(^1/_2)J_{AB}-F-(^1/_2)\delta_{AB}$	$2(d_1-d_2)^2$
16	$E_{-1}-2E_0$	$D_{AB}+(^1/_2)J_{AB}+F-(^1/_2)\delta_{AB}$	$2(d_1+d_2)^2$

where: $\delta_{AB} = \nu_B - \nu_A = -\nu_A$

$\Delta_1 = [(^3/_2)D_{BB} - D_{AB} - (^1/_2)J_{AB} - (^1/_2)\delta_{AB})^2 + (^3/_4)(D_{AB} - J_{AB})^2]^{1/2}$

$\Delta_2 = [(^3/_2)D_{BB} - D_{AB} - (^1/_2)J_{AB} + (^1/_2)\delta_{AB})^2 + (^3/_4)(D_{AB} - J_{AB})^2]^{1/2}$

$E = [(D_{AB} - J_{AB})^2 + (^1/_4)\delta_{AB}^2]^{1/2}$; $F = (^1/_2)[(D_{AB} - J_{AB})^2 + \delta_{AB}^2]^{1/2}$

$\dfrac{a_1}{a_2} = \dfrac{\delta_{AB} - 3D_{BB} + 2D_{AB} + J_{AB} - 2\Delta_1}{\sqrt{3}(D_{AB} - J_{AB})}$; $a_1^2 + a_2^2 = 1$; $\dfrac{b_1}{b_2} = -\dfrac{^1/_2\delta_{AB} + E}{D_{AB} - J_{AB}}$; $b_1^2 + b_2^2 = 1$

$\dfrac{c_1}{c_2} = \dfrac{\delta_{AB} + 3D_{BB} - 2D_{AB} - J_{AB} - 2\Delta_2}{\sqrt{3}(D_{AB} - J_{AB})}$; $c_1^2 + c_2^2 = 1$; $\dfrac{d_1}{d_2} = \dfrac{\delta_{AB} + 2F}{(D_{AB} - J_{AB})}$; $d_1^2 + d_2^2 = 1$

The spectra of 2,3,4,6- and 2,3,5,6-tetrachloroanisole have been studied at 60 MHz [44]. Fig. 9 shows the spectrum of the former with assigned lines. The parameters derived are presented in Table 12.

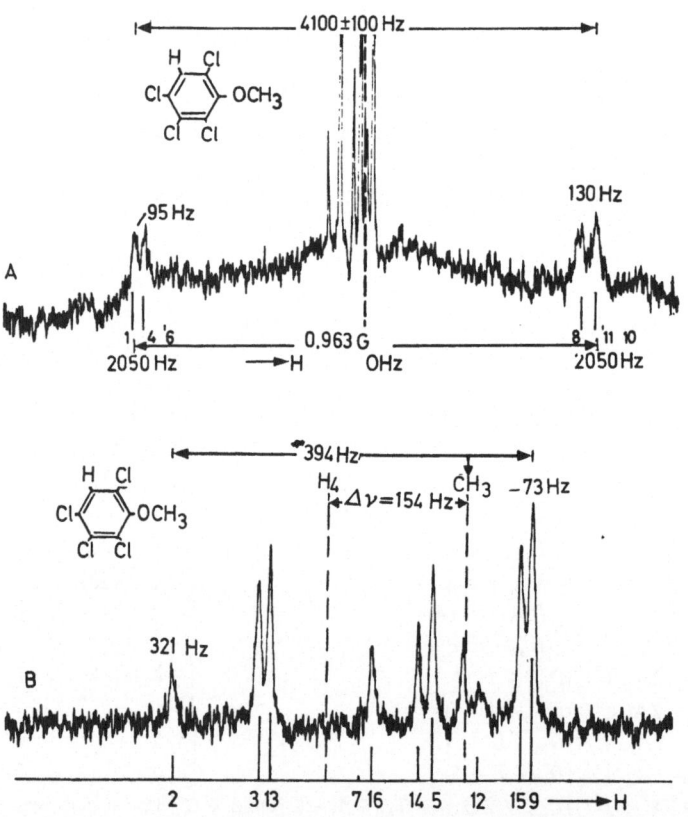

Fig. 9. Proton magnetic resonance spectrum of 2,3,4,6-tetrachloroanisole in the nematic phase of (IV) at 92°C [44]. The assignment of lines corresponds to Table 11. Spectrometer frequency = 60 MHz. (A) Complete spectrum. (B) Central part expanded. (Reprinted from Z. Naturforsch. 20a, 1401 (1965), copyright (1965). Reproduced by permission of the copyright owner)

Table 12. *Parameters obtained from the spectra of 2,3,4,6- and 2,3,5,6-tetrachloroanisole*

Compound	Structure	Temp. °C	Liquid crystal	Results	Reference
2,3,4,6-tetra-chloroanisole	OCH$_3$ (structure)	92	(IV)	$D_{BB} = 666\,\text{Hz}$ $D_{AB} = 56\,\text{Hz}$ $v_B - v_A = 154\,\text{Hz}$	[44]
2,3,5,6-tetra-chloroanisole	OCH$_3$ (structure)	74	(IV)	$D_{BB} = 574\,\text{Hz}$ $D_{AB} = -21.4\,\text{Hz}$ $v_B - v_A = -161.6\,\text{Hz}$	[44]

7.1.6. The System of Five Spins (A$_3$B$_2$)

Five spin systems of the type A$_3$B$_2$ are defined by the five parameters δ_{AB}, D_{AA}, D_{BB}, D_{AB} and J_{AB}. In the staggered conformations the monosubstituted ethanes belong to this class. They have C$_s$-symmetry and hence three parameters describe the orientation. The three direct couplings give information about the orientation only. The spectra of ethyl iodide in the nematic phase of (III) [41] and of ethyl alcohol [26] in the lyotropic mesophase have been studied. The former (Fig. 10) illustrates the concept of 'full equivalence' in NMR spectra of oriented molecules. In the analysis total spin sub-spectra may be identified. The parameters reproduced in Table 13 have been obtained from a modified version of the LAOCOON II programme.

Table 13. *Parameters obtained for the spectrum of ethyl iodide dissolved in the nematic phase of (III). $J_{AB} = 7.16\,Hz$ (assumed). Chemical shifts are relative to TMS.*

Temperature (°C)	Parameters							
	δ_A (ppm)	δ_B (ppm)	D_{AA} (Hz)	D_{BB} (Hz)	D_{AB} (Hz)	$c_{3z^2-r^2}$	$c_{x^2-y^2}$	c_{xy}
81	1.47	3.14	581.5	1349	-132.5	-0.1417	0.0592	0.1533
76	1.48	3.12	594.0	1390.5	-132.5	-0.1460	0.0597	0.1586
71	1.48	3.11	600.0	1417.0	-134.0	-0.1488	0.0597	0.1625

The relation between motional constants and the direct couplings are given in eqs. (31)–(33). The system of reference is defined as follows. The centre of gravity of the CH_3 group denotes the origin; the x-axis is taken along the $C-C$ bond, and the x and y coordinates of the iodine nucleus are positive.

$$D_{AA} = -2380c_{3z^2-r^2} + 4122.5c_{x^2-y^2} \tag{31}$$

$$D_{BB} = -9520.5c_{3z^2-r^2} \tag{32}$$

$$D_{AB} = 926.5c_{3z^2-r^2} - 1677.5c_{x^2-y^2} - 640c_{xy} \tag{33}$$

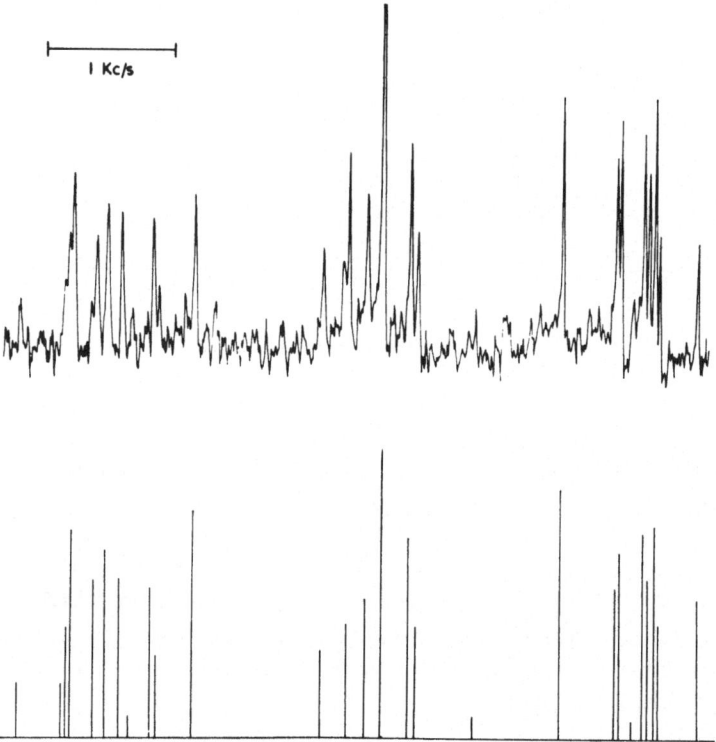

Fig. 10. NMR spectrum of ethyl iodide in the nematic phase of (III) [41]. Concentration = 20 mole %. Temperature = 71°C. Spectrometer frequency is 100 MHz. (Reprinted from Mol. Phys. 13, 365 (1967), copyright (1967). Reproduced by permission of the copyright owner)

The function $P(\theta, \varphi)$ describing the anisotropic motion at 71°C is given by eq. (34):

$$P(\theta, \varphi) = 0.092818 - 0.039722 \cos^2 \theta + 0.009197 \sin^2 \theta \cos 2\varphi \\ + 0.025050 \sin^2 \theta \sin 2\varphi \tag{34}$$

Another representation is obtained by plotting $P(\theta, \varphi)$ in the (x, y), (x, z) and (y, z) planes as shown in Fig. 11.

The molecule is oriented preferentially with its longest axis in the direction of the magnetic field.

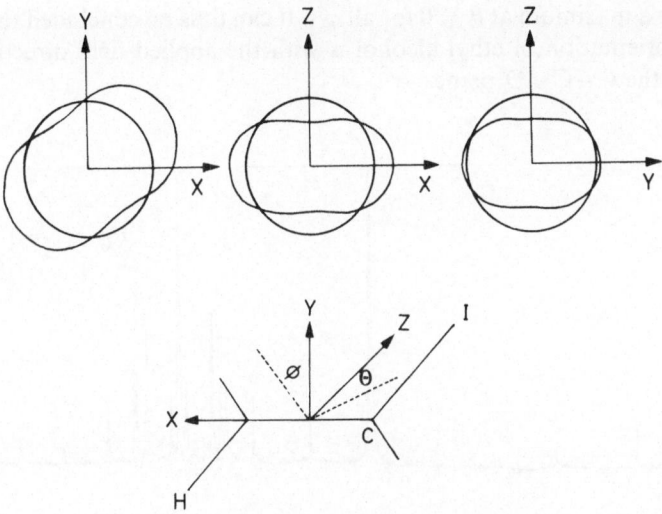

Fig. 11. Anisotropic motion of ethyl iodide in the nematic phase of (III) (20 mole % concentration; temperature 71°C)

The ethyl alcohol spectrum has been studied in a lyotropic solvent (Fig. 12). The nature of ordering in this phase allows rotation of the sample tube without destroying orientation and gives line-widths in the range of 1 – 4 Hz. The mesophase is relatively viscous and appears to orient relatively slowly. The highest quality spectra are obtained by spinning the sample in the magnetic field for several hours.

A modified version of LAOCN 3 has been used for the analysis. The results are summarised in Table 14.

Table 14. *Parameters for the spectrum of ethyl alcohol in the lyotropic mesophase. J_{AB} is assumed positive*

$v_A - v_B$	J_{AB}	D_{AA}	D_{AB}	D_{BB}	$c_{3z^2-r^2}$	$c_{x^2-y^2}$	c_{xy}
2.462 ± 0.11	7.05	-23.19	-31.30	-390.35	0.03988	0.01760	-0.05981
	± 0.07	± 0.07	$+0.06$	± 0.06			
(ppm)	(Hz)	(Hz)	(Hz)	(Hz)			

Eq. (35) describes the anisotropic motion:

$$P(\theta, \varphi) = 0.076029 + 0.010646 \cos^2 \theta + 0.002712 \sin^2 \theta \cos 2\varphi \\ - 0.009216 \sin^2 \theta \sin 2\varphi . \tag{35}$$

$P(\theta, \varphi)$ has a maximum at $\theta = 0$ for all φ's. It can thus be concluded that the most probable orientation of ethyl alcohol is with the applied field direction perpendicular to the $C - C - O$ plane.

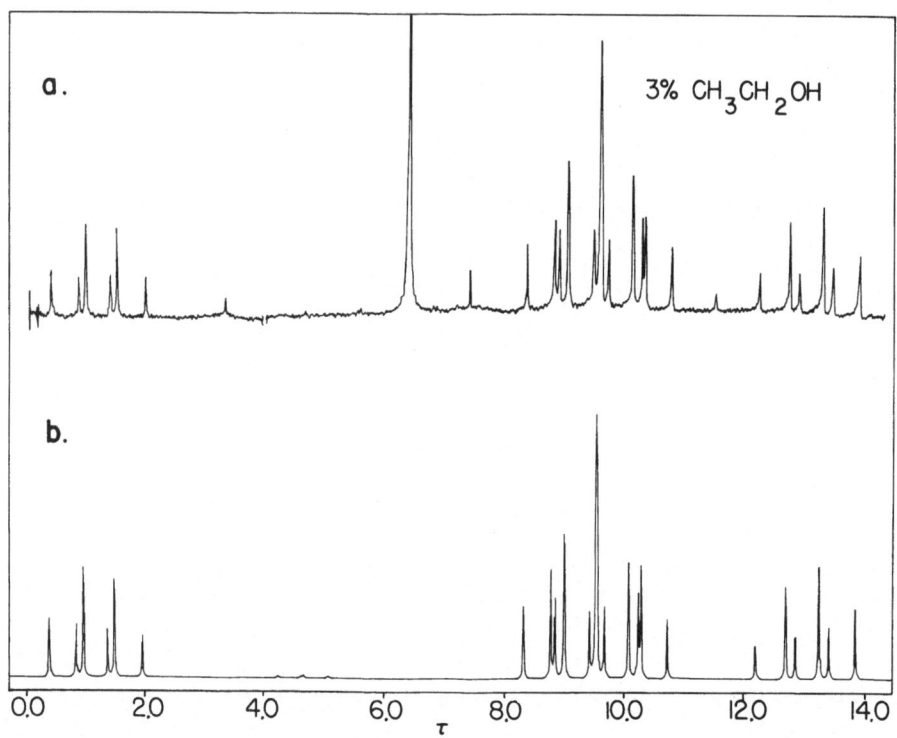

Fig. 12. PMR spectrum of ethyl alcohol in the lyotropic mesophase [26]. a) experimental, b) computer simulation. (Reprinted from Mol. Crist. Reproduced by permission of the copyright owner and the authors)

7.1.7. First Order Spectra of 7 and 10 Spins

First order proton spectra of oriented molecules (weak direct coupling) have been observed for molecules in the nematic phase of poly-γ-benzyl-L-glutamate [22] (PBLG). The orientation in such cases is weak. Methylene chloride, dimethylformamide, and p-xylene have been studied.

Fig. 13 shows a spectrum of p-xylene.

The triplet, centered at $+652\,Hz$ relative to the doublet of oriented methylene chloride, arises from the splitting of the three protons of each of the two methyl groups in p-xylene, whereas, the doublet at $-392\,Hz$ is due to direct coupling with ortho protons. Further direct coupling is too small to be observed. The absolute value of the orientation parameter for the $(H - H)_{ortho}$ axis is approximately 0.009.

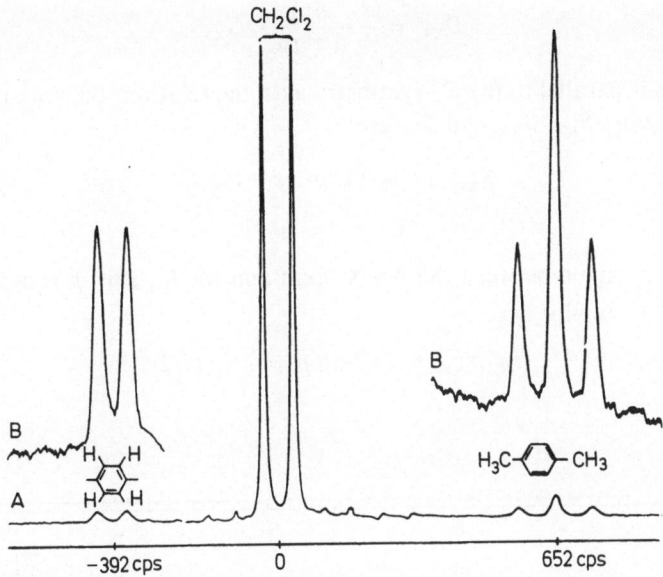

Fig. 13. NMR spectrum of 5% *p*-xylene in PBLG–CH_2Cl_2 at 220 MHz [22]. In B the spectrum of *p*-xylene is recorded at higher gain than in A. (Reprinted from J. Am. Chem. Soc. **90**, 3880 (1968), copyright (1968) by the American Chemical Society. Reprinted by permission of the copyright owner)

7.2. Systems Providing Information on the Geometry of Molecules

7.2.1. Systems of Four Spins (AB_3 and AX_3) with C_3-Symmetry

Spectra of the AB_3 or AX_3 systems with C_3-symmetry have undergone extensive study since they are of the simplest type which give information on the molecular geometry. A rotating methyl group together with either a proton or a heteronucleus situated on the C_3-axis of symmetry are examples of the two types.

The analysis of AB_3 type spectra has been discussed in section 7.1.5.

AX_3-type spectra can be easily analysed. The A-part consists of a 1:3:3:1 quartet with repeated spacing of $|J_{AX} + 2D_{AX}|$. The X_3-part is a doublet of 1:2:1 triplets (spacing = $|J_{AX} + 2D_{AX}|$ for the doublet and $3|D_{XX}|$ for the triplet). Relative signs of J and D must be known for a correct determination of D. Normally, the sign of J is taken from the isotropic spectrum whereas that of D is derived experimentally from similar molecules. Erroneous relative signs lead to a contradictory or strongly temperature and concentration dependent experimentally determined geometry [64].

There are two direct couplings but a single parameter fully describes orientation. The direct couplings D_{AX} and D_{XX} are given by eq. (36) and (37):

$$D_{AX} = -\frac{h\gamma_A\gamma_X}{4\pi^2 r_{AX}^3} \cdot S_{AX} \tag{36}$$

$$D_{XX} = -\frac{h\gamma_X^2}{4\pi^2 r_{XX}^3} \cdot S_{XX} \qquad (37)$$

If the z-axis is parallel to the C_3-symmetry axis, the relations between the orientation parameters S_{AX}, S_{XX} and S_{C_3} are:

$$S_{AX} = (^1/_2)(3\cos^2\beta - 1)S_{C_3} \qquad (38)$$

$$S_{XX} = -(^1/_2)S_{C_3} \qquad (39)$$

where β is the angle between the $A-X$ bond and the C_3-axis. β is dependent on the distances r_{XX} and r_{AX}:

$$(r_{XX}/r_{AX}) = \sqrt{3}\sin\beta = 2\sin(\alpha/2). \qquad (40)$$

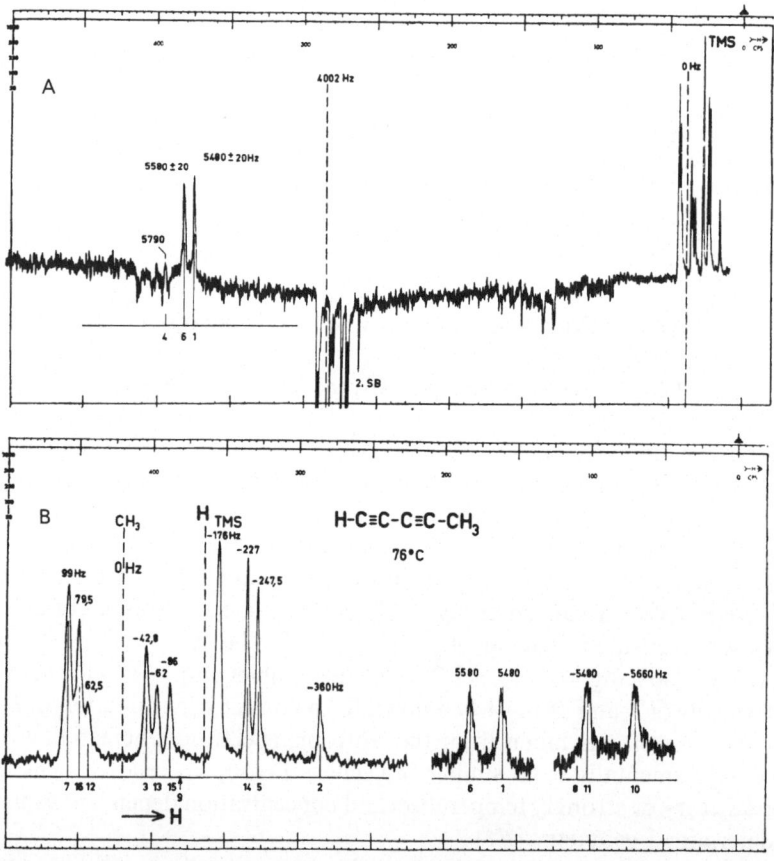

Fig. 14. Experimental and theoretical spectra of 1,3-pentadiyne in the nematic phase of (III) [57]. Concentration = 25 mole %. Temperature = 76°C. Spectrometer frequency = 100 MHz. (A) Central and low field regions. (B) Complete spectrum expanded. The line-numbers correspond to those of Table 11. (Reprinted from Mol. cryst. 1, 503, (1966), copyright (1966). Reproduced by permission of the copyright owner and the authors)

Eq. (40) also gives the correspondence between angle α (XCX or HCH in a methyl group for example) and β.

From eqs. (36 – 40) it follows that:

$$(D_{AX}/D_{XX}) = \left(\frac{\gamma_A}{\gamma_X}\right)\left(\frac{r_{XX}}{r_{AX}}\right)^3\left[\left(\frac{r_{XX}}{r_{AX}}\right)^2 - 2\right]. \tag{41}$$

Formula (41) is applicable to AB_3 cases as well. It shows that the ratio of inter-nuclear distances can be obtained from the values of the direct couplings. This ratio can further be used to obtain the angles α and β.

Typical AB_3 cases which have been studied are 1,3-pentadiyne [57], propyne [57, 45] and 2,3,5,6-tetrachlorotoluene [44]. The spectrum of pentadiyne is shown in Fig. 14.

Spectra of $^{13}CH_3CN$, $CH_3^{13}CN$, $CH_3C^{15}N$, $^{13}CH_3OH$, $^{13}CH_3I$ and $HC \equiv C - CF_3$ [57, 64, 66, 63] have been analysed as AX_3-type spectra. As an example, the spectrum of $^{13}CH_3CN + CH_3CN$ is shown in Fig. 15.

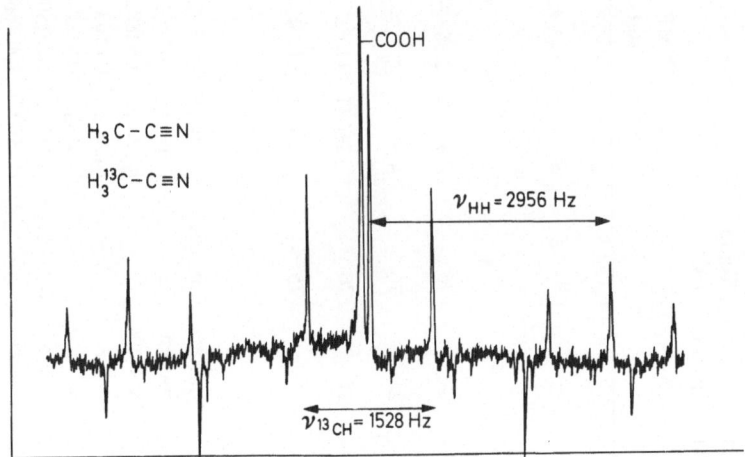

Fig. 15. PMR spectrum of a mixture of CH_3CN and $^{13}CH_3CN$ in the nematic phase of (VIII). The sample rotates at approximately 150 rev./min. [64]. Concentration = 24.5 mole %. Temperature = 96°C. Spectrometer frequency = 100 MHz. Signals with inverted intensity are 2(n) KHz side bands introduced by the integrator. The strongest line is due to the rapidly exchanging carboxylic proton of the solvent

Sub-spectra of the type AX_3 (Fig. 16) are observed in the AX_3Z spectrum of methyl isocyanide ($^{13}CH_3^{14}NC$) [65] where the interaction of the protons with the ^{14}N nucleus leads to a splitting.

Available information on such systems is summarised in Table 15. In some cases, Table 15 shows disagreement between geometrical parameters obtained from NMR and microwave spectroscopies. This may be due to molecular vibrations. Slight deviations between the data for the same solute in different solvents (Table 15) may arise from variations in intermolecular forces.

Table 15. *Molecular geometry as obtained from the spectra of types AB₃, AX₃, AX₃Z with C₃-symmetry*

Compound	Structure	Liquid crystal	Concentration (mole%)	Temp. (°C)	Parameter	NMR results	Geometry-information Microwave data	Reference
2,3,5,6-tetrachlorotoluene		(III)	–	82	R/r	3.25 ± 0.03	–	[44]
propyne	$HC{\equiv}C{-}CH_3$	(III)	–	75	R/r	2.353 ± 0.005	2.352 [67]	[57]
1,3-Pentadiyne	$HC{\equiv}C{-}C{\equiv}C{-}CH_3$	(III)	25	76	R/r	3.705 ± 0.03	–	[57]
methylalcohol	$^{13}CH_3OH$	(VIII)	15	70	$<HCH$	$110°3' \pm 8'$	$109°2' \pm 45'$ [69]	[57]
methyliodide	$^{13}CH_3I$	(III)	–	75	$<HCH$	$111°42' \pm 2'$	$111°25'$ [70]	[57]
methylcyanide	$H_3{}^{13}CCN$, $H_3C^{13}CN$, $H_3CC^{15}N$, etc.	(III)	various concentrations	(68–85)	$<HCH$ $<HCC$ r_{HH}/r_{CH} r_{HH}/r_{CCH} r_{HH}/r_{NH}	$109° (3.5 \pm 5)'$ $109°(52.9 \pm 0.5)'$ 1.62882 ± 0.00009 0.8563 ± 0.0009 0.5772 ± 0.0005	r_o [67] r_s [67] $109°16.5'$ $109°29.5'$ 0.85935 0.85765	[57, 64]
		(VIII)	several concentrations (15–35)	(86–110)	$<HCH$ $<HCC$ r_{HH}/r_{CH} r_{HH}/r_{NH}	$108°(59.5 \pm 0.5)'$ $109°(56.8 \pm 0.5)'$ 1.62815 ± 0.00017 0.5745 ± 0.0014	$109°40'$ $109°27'$ 1.6310 1.6332	

				Parameter	Value	Value	Ref
methyl-isocyanide[a] $^{13}CH_3{}^{14}NC$	3:3:1 mixture of (II, III + I)	various concentrations (15–35)	27	< HCH < HCC r_{HH}/r_{CH} r_{HH}/r_{NH}	109°(11.3 ± 1)' 109°(45.4 ± 1.2)' 1.6301 ± 0.0002 0.5727 ± 0.0006	 0.57255 0.5705	[65]
	(XII + XIII)		20 34	< α r_{CN}/r_{CH}	109°45' ± 3' 1.290 with negative J_{NH} 1.232 with positive J_{NH}	109°46' [68] r_0 r_s 1.304 [68] 1.291 [67]	
3,3,3-trifluoro-propyne HC≡C–CF$_3$	various liquid crystals, concentrations and temperatures			r_{HF}/r_{FF}	2.030 ± 0.025	2.038 ± 0.029	[66]

r is the distance between the methyl protons and R that between the methyl and the acetylenic or ring-protons.

[a] Recently the ratio r_{NH}/r_{HH} for this molecule has been found to be 1.144 ± 0.01 or 1.112 ± 0.01 for negative or positive J_{NH} respectively [141]. The microwave value is 1.146 [67]. HCH- and HHgC-angles in methylmercuric chloride have been determined to be (106.2 ± 0.6)° and (23.6 ± 0.1)° respectively [143].

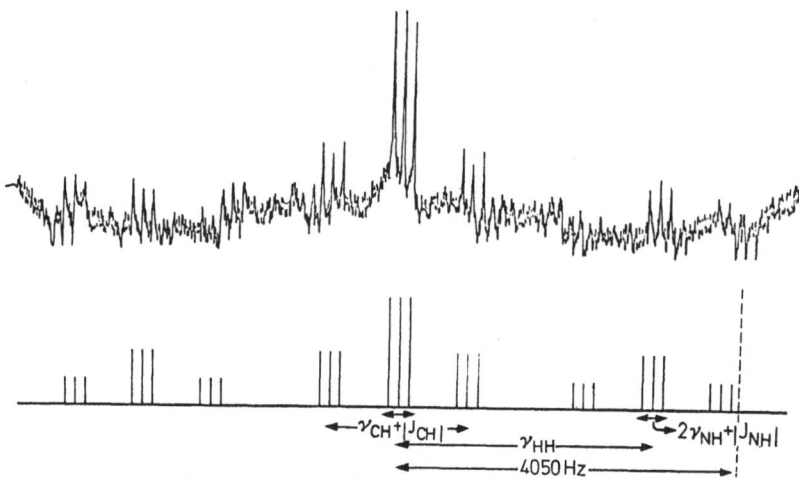

Fig. 16. Proton magnetic resonance spectrum of methyl isocyanide in the nematic phase of (XII + XIII) [65]. Concentration = 20 mole %. Temperature = 34°C. (Reprinted from Z. Naturforsch. 23a, 467 (1968), copyright (1968). Reproduced by permission of the copyright owner)

7.2.2. The System of Four Spins (AA′BB′) with C_{2v}-Symmetry

A planar four spin system with C_{2v}-symmetry has four different direct couplings and requires two parameters to describe the orientation (Fig. 17). The Hamiltonian of the system is given in Table 16.

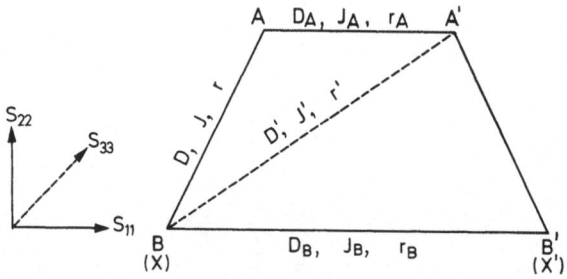

Fig. 17. Direct couplings and orientation parameters in a four spin system of the AA′BB′ (AA′XX′) type with C_{2v}-symmetry

Transition frequencies and intensities for a special case in which the chemical shift between the two groups and the indirect couplings are equal to zero, can be represented analytically as functions of the parameters (Table 17).
Table 17 may be used for preliminary analysis of an AA′BB′ spectrum. The final analysis can be performed with the aid of a computer, including indirect coupling constants and chemical shifts.

Table 16. *Hamiltonian matrix for oriented 4-spin systems with* C_{2v}-symmetry.
$\omega_3 = \omega_4 = 0$, $\omega_2 = \omega_5 = -\delta$. $(^1/_4)\,K$ *has been subtracted from all the diagonal elements**

No.	Diagonal elements	Off-diagonal elements
1	$-\delta + (^1/_2)N + (^1/_2)Q + S$	
2	$-(^1/_4)(Q - 3R)$	
3	$-\delta - (^1/_4)(Q + 3R)$	$\mathscr{H}_{23} = (^1/_2)(N - S)$
4	$+\delta + (^1/_2)Q - (^1/_2)N - S$	$\mathscr{H}_{45} = +(^1/_2)(N - S)$
5	$-Q$	$\mathscr{H}_{56} = +(^1/_2)(N - S)$
6	$-\delta + (^1/_2)Q - (^1/_2)N - S$	$\mathscr{H}_{67} = -(^1/_2)(L - T)$
7	$-K$	$\mathscr{H}_{46} = 0$
		$\mathscr{H}_{57} = +(^1/_2)(L + 2T)$
		$\mathscr{H}_{47} = -(^1/_2)(L - T)$
8	$-(^1/_4)(Q - 3R)$	
9	$+\delta - (^1/_4)(Q + 3R)$	$\mathscr{H}_{89} = +(^1/_2)(N - S)$
10	$+\delta + (^1/_2)N + (^1/_2)Q + S$	
11	$+(^1/_2)M + (^1/_4)R - (^1/_2)K + (^1/_4)Q$	
12	$-\delta - (^1/_2)M - (^1/_4)R - (^1/_2)K + (^1/_4)Q$	$\mathscr{H}_{11\,12} = +(^1/_2)(L - T)$
13	$-(^1/_2)K - (^1/_2)Q + (^1/_2)M - (^1/_2)R$	
14	$-(^1/_2)K - (^1/_2)Q - (^1/_2)M + (^1/_2)R$	$\mathscr{H}_{13\,14} = (^1/_2)(L + 2T)$
15	$+(^1/_2)M + (^1/_4)R - (^1/_2)K + (^1/_4)Q$	
16	$\delta - (^1/_2)M - (^1/_4)R - (^1/_2)K + (^1/_4)Q$	$\mathscr{H}_{15\,16} = (^1/_2)(L - T)$

* $K = J_A + J_B$. $M = J_A - J_B$. $N = J + J'$. $L = J - J'$. $J_A = J_{34}$. $J_B = J_{25}$. $J = J_{23} = J_{45}$. $J' = J_{24} = J_{35}$. $Q = D_A + D_B$; $D_A = D_{34}$. $R = D_A - D_B$; $D_B = D_{25}$. $S = D + D'$; $D = D_{23} = D_{45}$. $T = D - D'$; $D' = D_{24} = D_{35}$.

Spectra of furan, thiophene, benzofurazan oxide, symmetrical ortho di-substituted benzenes and pyridazine [50, 71, 72, 73, 124] have been investigated. The spectra of the first four contain a maximum of 12 intense lines, of which numbers 3 and 4 and/or 9 and 10 (Fig. 18) may overlap. Other transitions are too weak to be detected without the help of a computer of average transients.

In the pyridazine molecule 19 lines are observed [124].

The results of the analyses are summarised in Table 18.

Relations between the ratios of the inter-proton distances and the direct couplings are given in eqs. (42 – 44). They are valid irrespective of the molecular orientation. The values derived for the molecules studied are included in Table 18.

$$(r_B/r_A) = (D_A/D_B)^{1/3} \tag{42}$$

$$D(r/r_A)^5 - D'\{(D_A/D_B)^{1/3} + (r/r_A)^2\}^{5/2} = -D_A(D_A/D_B)^{1/3} \tag{43}$$

$$(r'/r_A)^2 = (D_A/D_B)^{1/3} + (r/r_A)^2 \tag{44}$$

Table 17. *Transition frequencies and intensities for an oriented 4-spin system with C_{2v}-symmetry, without chemical shift and indirect couplings. Lines with zero intensity are not included. The table gives only half the spectrum. The other half is symmetrical to it* *

Transition	Frequency	Intensity
$1s_1 - s_2$	$(^3/_4)(D_A + D_B) + (D + D') - (^1/_2)R_1$	$2(1 - Q')^2/(1 + Q'^2)$
$2s_1 - s_2$	$(^3/_4)(D_A + D_B) + (D + D') + (^1/_2)R_1$	$2(1 + Q')^2/(1 + Q'^2)$
$2s_0 - 1s_1$	$-(^3/_4)(D_A + D_B) + (D + D') + (^1/_2)R_1$	$(1 + Q')^2/(1 + Q'^2)$
$3s_0 - 1s_1$	$(^1/_2)(D + D') + (^1/_2)(R_1 - R_2)$	I_1
$4s_0 - 1s_1$	$(^1/_2)(D + D') + (^1/_2)(R_1 + R_2)$	I_2
$2s_0 - 2s_1$	$-(^3/_4)(D_A + D_B) + (D + D') - (^1/_2)R_1$	$(1 - Q')^2/(1 + Q'^2)$
$3s_0 - 2s_1$	$(^1/_2)(D + D') - (^1/_2)(R_1 + R_2)$	I_3
$4s_0 - 2s_1$	$(^1/_2)(D + D') + (^1/_2)(R_2 - R_1)$	I_4
$1a_0 - 2a_1$	$(^3/_4)(D_A + D_B) - (^3/_2)R_3$	2
$2a_0 - 1a_1$	$(^3/_4)(D_A + D_B) + (^3/_2)R_3$	2

* $R_1 = [(^9/_4)(D_A - D_B)^2 + (D + D')^2]^{1/2}$
$R_2 [(^9/_4)(D_A + D_B)^2 + 3(D + D')^2 + 6(D - D')^2 - 3(D_A + D_B) \cdot (D + D')]^{1/2}$
$R_3 [(^1/_4)(D_A - D_B)^2 + (D - D')^2]^{1/2}$
$Q' = (D + D')/[(^3/_2)(D_A - D_B) + R_1]$.

The expressions for intensities I_1, I_2, I_3 and I_4 are too tedious to be given. However, they fulfill the relations: $I_1/I_3 = I_2/I_4 = [(1 - Q')/(1 + Q')]^2$.

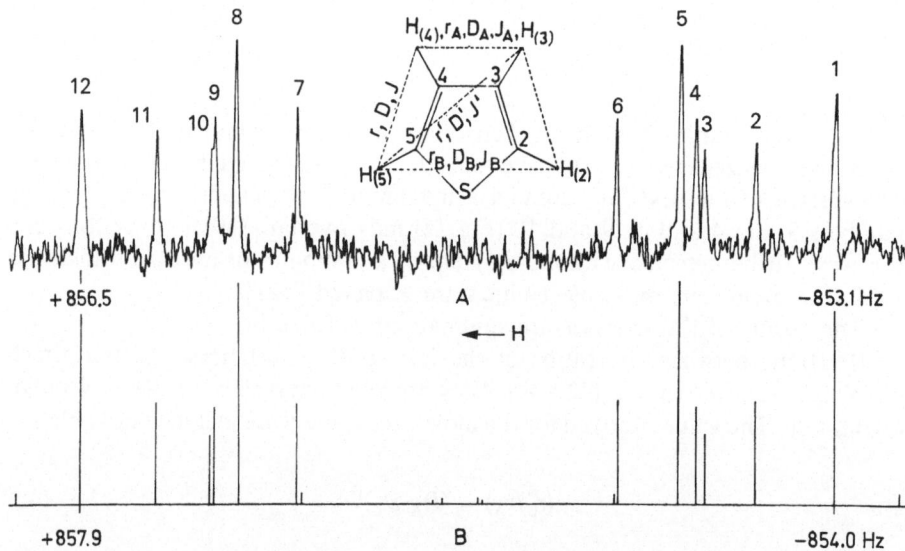

Fig. 18. Observed (A) and calculated (B) NMR spectra of thiophene in the nematic phase of (III) [50]. Concentration = 23 mole %. Temperature = 80°C. Spectrometer frequency = 60 MHz

Table 18. *Information from the AA'BB' spectra of molecules studied in the nematic phase*

Information	Compound						
	Thiophene	Furan	Benzofurazan oxide	o-Dibromo-benzene	o-Dichloro-benzene	o-Dicyano-benzene	Pyridazine
liquid crystal	(III)	(II)	(II)	(III)	(II)	(II)	(40% XI + 60% XII)
concentration (mole %)	23	20	19	20	15	14	26
temperature (°C)	80	70	56	79	65	65	60
indirect couplings (Hz)	$J_A = 3.3$, $J_B = 2.8$, $J' = 4.7$, $J'' = 1.0$	$J_A = 3.4$, $J_B = 1.6$, $J = 1.8$, $J' = 0.8$	$J_A = 6.5$, $J_B = 1.2$, $J = 9.3$, $J' = 0.9$	$J_A = 7.4$, $J_B = 0.3$, $J = 8.1$, $J' = 1.6$			$J_A = 8.0$, $J_B = 1.4$, $J = 5.05$, $J' = 1.85$
D_A (Hz)	-175.3	-86.3	-290.1	-302.5	-289.5	-345.6	-476.5
D_B (Hz)	-33.0	-27.4	-33.0	-38.7	-36.1	-43.6	-70.6
D (Hz)	-394.4	-275.2	-604.6	-746.2	-692.2	-919.8	-313.1
D' (Hz)	-55.8	-41.0	-75.9	-87.0	-83.0	-102.7	-86.0
$\nu_A - \nu_B$ (Hz)	10	68	20.3	-1.3	6.3	1.0	189.9
r_B/r_A	1.745±0.010 (1.719)	1.47±0.02 (1.485)	2.06±0.04	1.985±0.014	2.002±0.014	1.944±0.010	1.890±0.004
r/r_A	0.995±0.005 (0.998)	0.98±0.02 (0.994)	1.03±0.03	0.995±0.010	1.017±0.010	0.991±0.010	0.988±0.010
r'/r_A	1.653±0.006 (1.648)	1.56±0.02 (1.571)	1.77±0.04	1.725±0.007	1.742±0.007	1.725±0.007	1.693±0.007
S_{11}	0.0265 ±0.0003	0.0151 ±0.0004	0.0369	0.0384 ±0.0002	0.0368 ±0.0002	0.0439 ±0.0002	0.0644 ±0.0019
S_{22}	0.0639 ±0.0002	0.0469 ±0.0003	0.1004	0.1130 ±0.0001	0.1050 ±0.0001	0.1410 ±0.0001	0.0347 ±0.0006
S_{33}	-0.0904	-0.0620	-0.1373	-0.1514	-0.1418	-0.1849	-0.0991
S_{22}/S_{11}	2.4113	3.1060	2.7209	2.9427	2.8533	3.2118	0.5388
reference	[50]	[71]	[73]	[72]	[72]	[72]	[124]

The microwave values of the distance ratios are given in parentheses.

4*

The orientation parameters S_{11} and S_{22} are related to the direct couplings and the inter-proton distances by (45) and (46):

$$S_{11} = -\left(\frac{4\pi^2}{h\gamma^2}\right) D_A r_A^3 \tag{45}$$

$$S_{22} = -\left(\frac{4\pi^2}{h\gamma^2}\right) r^3 \left[D - (D_A/4)(r/r_A)^{-5}\{(D_A/D_B)^{1/3} - 1\}^2\right] \tag{46}$$

$$\cdot \left[1 - (^1/_4)(r/r_A)^{-2}\{(D_A/D_B)^{1/3} - 1\}^2\right]^{-1}.$$

S_{11} and S_{22} can be determined only if a proton-proton distance is known. The orientation parameters reported in Table 18 have been deduced from the r_A-values given in the literature. On the other hand S_{22}/S_{11} depends only on the ratios of the inter-proton distances and the direct couplings and hence can be derived from the NMR work alone (Table 18).

The fact that S_{22} is larger than S_{11} indicates the preferential orientation of the C_2-symmetry axis in the direction of the magnetic field. Pyridazine is an exception.

7.2.3. The System AA'A''A''' with D_{2d}-Symmetry

An AA'A''A''' system is a special case of the AA'BB' type in which $\delta_{AB} = 0$. The spectrum of allene [75, 76] which belongs to this class has been studied and the ratio of the inter-proton distances found to be 0.4778.

7.2.4. The System AA'XX' with C_{2v}-Symmetry

The Hamiltonian of the AA'XX'-spin system decays into submatrices of orders (1×1) and (2×2) due to weak coupling and C_2-symmetry. For the spectroscopy in isotropic solvents it has been shown by subspectral transformation [46] that these correspond to two A_2- and two AB-type sub-spectra with effective chemical shifts and effective coupling constants.

For the spectrum of an oriented AA'XX'-system, sub-spectral transformation does not allow to derive all the necessary relations, because the invariants of the Hamiltonian provide only two equations for the three parameters [39]. By inclusion of the intensity ratio, however, it is possible to demonstrate [60] that the AA' part of the spectrum of an oriented AA'XX'-system is a superposition of two sub-spectra of the type A_2 (oriented) with the effective Lamor frequency and direct coupling given by

$$\nu_a = \nu_A \pm (^1/_2)(J_{AX} + J_{AX'}) + (D_{AX} + D_{AX'}) \tag{46a}$$
$$D_{aa'} = D_{AA'}$$

and of two AB(oriented)-type sub-spectra with

$$\nu_a = \nu_A + (^1/_2)(J_{AX} - J_{AX'}) + (D_{AX} - D_{AX'})$$
$$\nu_b = \nu_A - (^1/_2)(J_{AX} - J_{AX'}) - (D_{AX} - D_{AX'}) \tag{46b}$$

and

$$J_{ab} = (J_{AA'} + J_{XX'}) - D_{XX'}$$
$$D_{ab} = D_{AA'}$$

(46c)

or

$$J_{ab} = (J_{AA'} - J_{XX'}) + D_{XX'}$$
$$D_{ab} = D_{AA'}$$

(46d)

By exchanging X and X' for A and A' in eqs. (46a – 46d) the corresponding relations are obtained for the sub-spectra of the XX'-part. Unlike the isotropic case, the oriented AA'XX' spectra have non-identical AA'- and XX'-parts.

^{1}H and ^{19}F spectra of 1,1-difluoroethylene have been investigated [77, 78] and the structural information obtained [77] is included in Table 19 together with the microwave data [79].

Table 19. *Structural information obtained from NMR spectra of 1,1-difluoroethylene in the nematic phase* [77]. *Symbols refer to Fig. 17*

Distance ratio	NMR value	Microwave data
r_B/r_A	1.017	1.1305
r/r_A	1.380	1.463
r'/r_A	1.686	1.519

A comparison between the NMR and the microwave values of inter-nuclear distances (Table 19) shows poor agreement. The deviations have been attributed to an anisotropic contribution in the indirect F–F coupling constant [77]. Provided that there is no anisotropic contribution to the H–H and H–F couplings, the magnitude of D_{FF}^{ind} is 18.1 Hz [77].

7.2.5. The System AA'A''A''' with D_{2h}-Symmetry

Four protons forming a rectangle represent a system of the AA'A''A''' type with D_{2h}-symmetry. This system is a special case of the AA'BB' type with C_{2v}-symmetry. Under the assumption that indirect couplings are negligible, the expressions for the frequencies and intensities of the allowed transitions can be derived from Table 17 by equating D_A to D_B. The general expressions including the indirect couplings have been reported in the literature [48, 80].

An AA'A''A''' (D_{2h}-symmetry) spectrum generally consists of five doublets placed symmetrically with respect to the centre. The intensities are in the ratios 2:3:1:1:1. Each component of the most intense doublet is further split due to indirect couplings. For protons this splitting is normally too small to be observed. All other lines are independent of the indirect couplings.

The spectra of symmetrical p-disubstituted benzenes [35, 48], p-benzoquinone [74], pyrazine [58, 74], ethylene [76, 81], tetrafluoroethylene [77], cyclobutadiene

Table 20. *Values of the parameters obtained in AA'A"A''' systems with D_{2h}-symmetry*

Compound	Structure	Liquid crystal	Temp. (°C)	Concentration (mole %)	D_{12} (Hz)	D_{14} (Hz)	D_{13} (Hz)	r_{14}/r_{12}	α (defined in column 2)	Reference
p-di-chloro-benzene		(III)	70	40	-2271 ± 2	137.5 ± 1.0	-6.4 ± 0.5	1.721 ± 0.005	120°23'	[48]
p-di-bromo-benzene		(III)	80	40	-2272 ± 2.5	149.5 ± 2	0.0 ± 1.0	1.723 ± 0.010	120°17'	[48]
p-diiodo-benzene		(III)	76	40	-2513 ± 1.5	170.5 ± 1.5	8.5 ± 1.0	1.751 ± 0.009	119°24'	[48]
ethylene		(III)	83	15	∓ 240.0	± 370.3	± 99.8	1.320 ± 0.006	—	[81]

tetra-fluoro-ethylene	$F(1)$, $F(4)$, $C=C$, $F(2)$, $F(3)$	(40% XII + 60% XIII)	20	609.5	764.8	268.8	1.194	—	[77]
p-benzo-quinone		(III)	82	13	−1578	+8	−44	1.745±0.015	122°30'±20' [74]
pyrazine		(III)	80	30	−317.5±1.0	−86.0±1.0	−50.0±1.0	—	[58]
pyrazine		(I)	27	9	−378.8±0.7	−30.0±1.5	−30.0±1.5	—	[58]
pyrazine		(II)	55	15.4	−561.7±1.0	−17.2±1.0	−28.4±1.0	1.66±0.02	120°22' [58]

P. Diehl and C. L. Khetrapal

Table 20 (continued)

Compound	Structure	Liquid crystal	Temp. (°C)	Concentration (mole %)	D_{12} (Hz)	D_{14} (Hz)	D_{13} (Hz)	r_{14}/r_{12}	α (defined in column 2)	Reference
cyclobutadiene iron tricarbonyl		(III)	76	25	437.5 ± 4.2	440.6 ± 4.2	162.2 ± 4.4	0.9977 ± 0.0045	$\approx 135°$	[80]
ethylene oxide		—	—	—	—	—	—	1.360	—	[76]
ethylene sulphide		—	—	—	—	—	—	1.349	—	[76]

irontricarbonyl [80], ethylene oxide and sulphide [76] have been studied. Fig. 19 reproduces one half of the symmetrical spectrum of *p*-dichlorobenzene. Table 20 contains the values of the parameters.

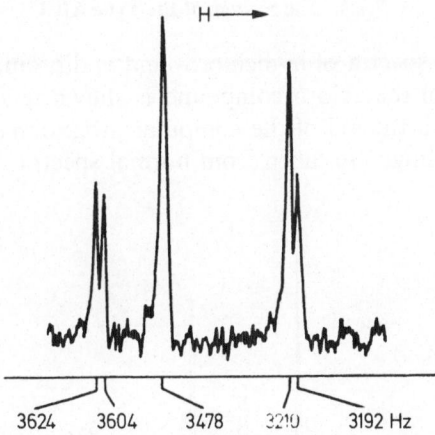

Fig. 19. NMR spectrum of *p*-dichlorobenzene in the nematic phase of (III) [48]. Concentration = 40 mole %. Temperature = 70° C. Spectrometer frequency = 56.4 MHz. (One half of the symmetrical spectrum is shown). (Reprinted from Rec. Trav. Chim. **87**, 417 (1968), copyright (1968). Reproduced by permission of the copyright owner and the authors)

In this system the ratio (r_{14}/r_{12}) is related to the direct couplings by (47):

$$D_{14}(r_{14}/r_{12})^5 - D_{13}\{1 + (r_{14}/r_{12})^2\}^{5/2} + D_{12} = 0. \tag{47}$$

Numerical results obtained for the studied compounds are included in Table 20.

On the basis of known CH and CC bond lengths, the angle α (column 2 of Table 20) can be calculated. In the *p*-disubstituted benzenes, the angle decreases with the size of the substituent.

It should be noted that the precision of determination of (r_{14}/r_{12}) depends not only upon the accuracy to which the D-values can be measured, but is also substantially influenced by their relative magnitudes. For pyrazine, for example, in the liquid crystal (III), eq. (47) has no solution in the positive range of (r_{14}/r_{12}), the curve passing through a minimum at $(r_{14}/r_{12}) = 1.53$ [58]. If the values of the direct couplings are modified by +1 Hz (accuracy of the measurement), two solutions are obtained at 1.391 and 1.755. A change in the opposite direction by the same amount does not provide a solution for positive (r_{14}/r_{12}). This introduces a large uncertainty in the ratio.

In the liquid crystal (I) on the other hand, two lines of relative intensity 1 are overlapping. This corresponds to $D_{13} \approx D_{14}$ within the limits of resolution and results in a lower precision.

The liquid crystal (II) as the solvent provides a reasonably accurate value of (r_{14}/r_{12}). The corresponding angle α is included in Table 20.

In conclusion it should be emphasised that there are spectra of oriented molecules which should in principle provide geometrical information, but fail to do so in practice.

7.2.6. The System of the Type AB₂C

In this class, the spectra of *m*-dichloro- and *m*-dibromo-benzenes have been studied [59]. That of the chloro compound is shown in Fig. 20. Analyses have been carried out with the aid of the computer programme Laocoonor. Values of the indirect couplings are taken from normal spectra. Table 21 presents the quantities obtained.

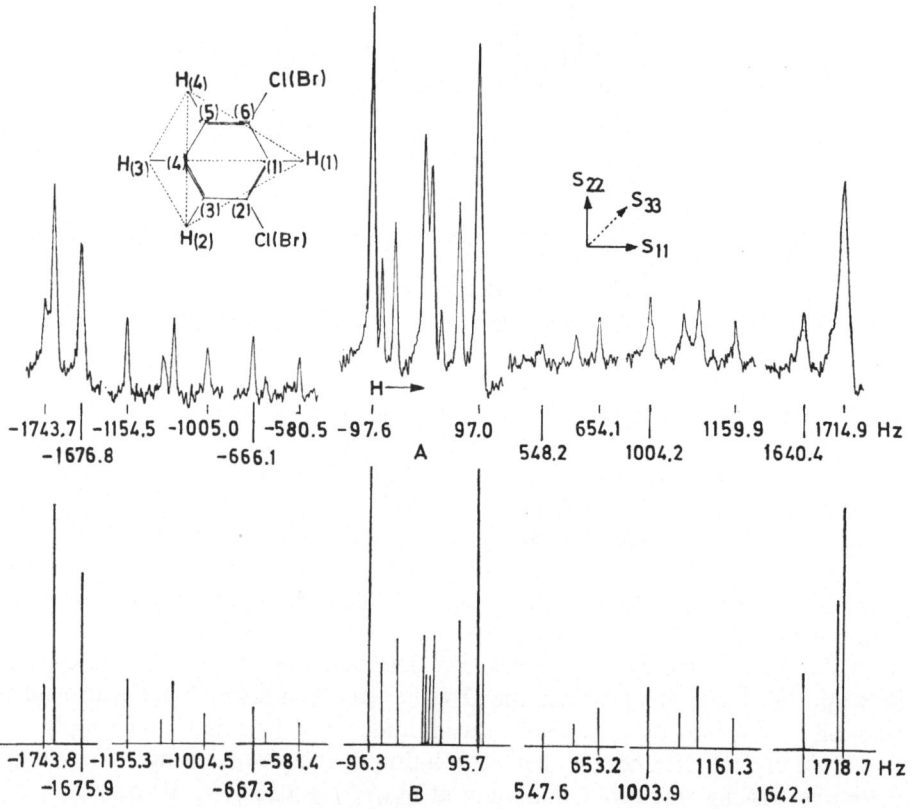

Fig. 20. (A) observed (B) calculated NMR spectra of *m*-dichlorobenzene in the nematic phase of (I) [59]. Concentration = 13.4 mole %, temperature = 27°C. Spectrometer frequency = 60 MHz

These compounds are planar with C_{2v}-symmetry. Each of the molecules has four different direct couplings related to the ratios of the inter-proton distances by eqs. (48), (49) and (50):

$$D_{14}\left[\tfrac{1}{4}y^2 + \left\{\left(\frac{D_{23}-\tfrac{1}{4}y^5 D_{24}}{(1-\tfrac{1}{4}y^2)D_{13}}\right)^{\tfrac{1}{3}} - (1-\tfrac{1}{4}y^2)^{1/2}\right\}^{1/2}\right]^{5/2} - \tfrac{1}{4}y^5 D_{24}$$

$$-\left(\frac{D_{23}-\tfrac{1}{4}y^5 D_{24}}{1-\tfrac{1}{4}y^2}\right)\left[\left(\frac{D_{23}-\tfrac{1}{4}y^5 D_{24}}{(1-\tfrac{1}{4}y^2)D_{13}}\right)^{\tfrac{1}{3}} - (1-\tfrac{1}{4}y^2)^{1/2}\right]^{2} = 0 \tag{48}$$

$$z = \left(\frac{D_{23}-\tfrac{1}{4}y^5 D_{24}}{(1-\tfrac{1}{4}y^2)D_{13}}\right)^{\tfrac{1}{3}} \tag{49}$$

$$x = \left[\tfrac{1}{4}y^2 + \left\{z-(1-\tfrac{1}{4}y^2)^{1/2}\right\}^2\right]^{1/2} \tag{50}$$

where $y = r_{24}/r_{23}$, $z = r_{13}/r_{23}$ and $x = r_{12}/r_{23}$ (Fig. 20).

Table 21. *Results of spectral analysis for meta dichloro- and meta dibromobenzene. Notations are given in Fig. 20*

	Compound	
	Meta dichlorobenzene	Meta dibromobenzene
liquid crystal	(I)	(I)
concentration (mole %)	13.4	12.0
$D_{12}(=D_{14})$(Hz)	-46.22 ± 0.14	-45.32 ± 0.17
$D_{23}(=D_{34})$(Hz)	-693.55 ± 0.11	-638.14 ± 0.13
D_{13} (Hz)	-2.75 ± 0.35	-2.78 ± 0.37
D_{24} (Hz)	-176.55 ± 0.23	-161.20 ± 0.29
$v_1 - v_2$ (Hz)	23.02 ± 0.40	23.33 ± 0.50
$v_3 - v_2$ (Hz)	9.36 ± 0.67	24.83 ± 0.74
r_{24}/r_{23}	1.7320 ± 0.0009	1.7346 ± 0.0009
$<H_{(2)}C_{(3)}C_{(4)}$	$119° 57' \pm 14'$	$119° 15' \pm 14'$
S_{11}	0.0028	0.0027
S_{22}	0.1165 ± 0.0009	0.1052 ± 0.0009
S_{33}	-0.1193	-0.1079

Eq. (48) is very sensitive even to small changes of y (r_{24}/r_{23}) and hence the value can be estimated quite precisely (Table 21). The result for y and the assumed values of the CC and CH bond lengths (1.395 Å and 1.084 Å respectively), have been used to determine the angle $H_{(2)}C_{(3)}C_{(4)}$ reproduced in Table 21. The angle is smaller for the bromo than for the chloro compound.

The magnitude of z can be determined from eq. (49) if y is known. However, the numerator of eq. (49) is the difference of two large quantities of nearly the same magnitude and the denominator contains the factor D_{13} which is small and hence relatively inaccurate ($D_{13} = -2.75 + 0.35$ Hz and -2.78 ± 0.37 Hz for the chloro- and the bromo-compound respectively). Consequently, the precision of z is low ($z = 2.01 \pm 0.32$ and 1.98 ± 0.37 for the two compounds respectively).

From y and z, x can be calculated (eq. 50). A value of 1.74 $\pm$ 0.28 has been obtained for the chloro compound and 1.71 $\pm$ 0.32 for the bromo.

Again, in these compounds the direct and indirect couplings are of opposite sign. Since the indirect couplings are known to be positive, the direct ones must be negative, indicating a preferential orientation of the molecular plane in the direction of the magnetic field. The larger value of S_{22} (Table 21) illustrates the preferred orientation of the C_2-symmetry axis perpendicular to the magnetic field.

7.2.7. The Five Spin System (AA′BB′C)

The proton spectrum of pyridine in the nematic phase (Fig. 21) is of the type AA′BB′C [55].

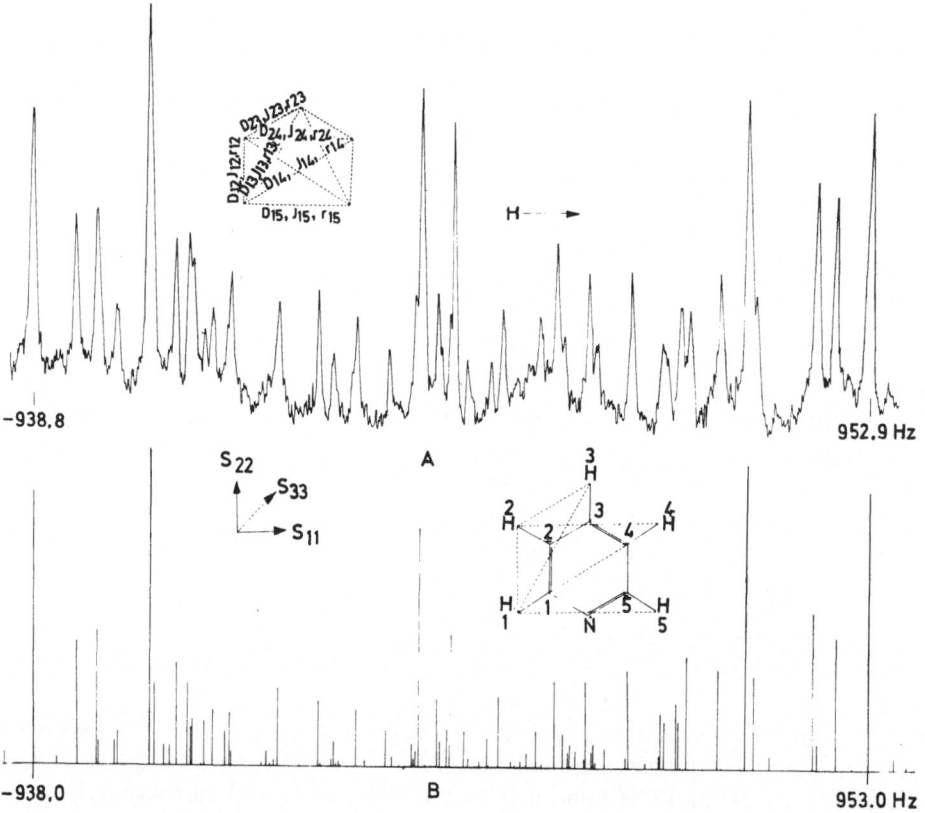

Fig. 21. (A) observed (B) calculated NMR spectra of pyridine oriented in the nematic phase of (II) [55]. Concentration = 26.4 mole %, temperature = 56°C. Spectrometer frequency = 60 MHz

It has been analysed iteratively with the LAOCOONOR programme. Values of the indirect coupling constants used have been taken from the literature [82]. The parameters obtained are summarised in Table 22.

Table 22. *The direct couplings, the ratios of the various inter-proton distances and the orientation parameters in pyridine (notations are explained in Fig. 21). ($J_{12} = 4.9\,Hz$; $J_{13} = 1.9\,Hz$; $J_{14} = 1.0\,Hz$; $J_{15} = -0.1\,Hz$; $J_{23} = 7.7\,Hz$; $J_{24} = 1.3\,Hz$ [82]).*

Parameter	Value (Hz)	Parameter	Value	
			Microwave [83]	NMR
$v_2 - v_1$ [a]	81.2 ± 0.4	r_{15}/r_{24}	0.962	0.962 ± 0.003
$v_3 - v_1$	59.1 ± 0.4	r_{12}/r_{24}	0.581	0.579 ± 0.003
D_{12}	-374.6 ± 0.1	r_{14}/r_{24}	1.140	1.139 ± 0.003
D_{13}	-65.0 ± 0.2	r_{23}/r_{24}	0.583	0.587 ± 0.004
D_{14}	-33.1 ± 0.1	r_{13}/r_{24}	1.003	1.009 ± 0.005
D_{15}	-45.8 ± 0.3	S_{11}	–	0.0265 ± 0.0003
D_{23}	-247.2 ± 0.2	S_{22}	–	0.0484 ± 0.0002
D_{24}	-40.8 ± 0.3	S_{33}	–	-0.0749

[a] $v_i - v_j$ = chemical shift of the nucleus i relative to that of j.

Pyridine has a planar proton skeleton with C_{2v}-symmetry. Ratios of the various inter-proton distances are related to the direct couplings by eqs. (51) to (55):

$$(r_{15}/r_{24}) = (D_{24}/D_{15})^{1/3} \tag{51}$$

$$D_{12}(r_{12}/r_{24})^5 - D_{14}\{(D_{24}/D_{15})^{1/3} + (r_{12}/r_{24})^2\}^{5/2} = -D_{24}(D_{24}/D_{15})^{1/3} \tag{52}$$

$$(r_{14}/r_{24})^2 = (D_{24}/D_{15})^{1/3} + (r_{12}/r_{24})^2 \tag{53}$$

$$\frac{[4D_{13}(r_{13}/r_{24})^5 - D_{24}(D_{24}/D_{15})^{2/3}]}{[4(r_{13}/r_{24})^2 - (D_{24}/D_{15})^{2/3}]} = \frac{[4D_{23}(r_{23}/r_{24})^5 - D_{24}]}{[4(r_{23}/r_{24})^2 - 1]} \tag{54}$$

and

$$4(r_{13}/r_{24})^2 = (D_{24}/D_{15})^{2/3} + \{[4r_{23}/r_{24})^2 - 1]^{1/2} + [4(r_{12}/r_{24})^2 - ((D_{24}/D_{15})^{1/3} - 1)^2]^{1/2}\}^2 . \tag{55}$$

Numerical results for pyridine are compared with the microwave data [83] in Table 22. The agreement is satisfactory.

Eqs. (56) and (57) relate the orientation parameters (S_{11} and S_{22}) to the inter-proton distances and the direct couplings.

$$S_{11} = -\left(\frac{4\pi^2}{h\gamma^2}\right) D_{24} r_{24}^3 , \tag{56}$$

$$S_{22} = -\left(\frac{4\pi^2}{h\gamma^2}\right)\left\{\frac{[4(r_{23}/r_{24})^5 D_{23} - D_{24}] r_{24}^3}{[4(r_{23}/r_{24})^2 - 1]}\right\} \tag{57}$$

S_{11} and S_{22} obtained for pyridine are included in Table 22 ($r_{24} = 4.274\,Å$). The positive signs of S_{11} and S_{22} are in agreement with the preferred orientation of the molecule with its plane in the direction of the magnetic field. S_{22} which is larger than S_{11} shows that the C_2-symmetry axis is preferentially oriented in the direction of the magnetic field. A similar observation has been made in symmetrical ortho disubstituted benzenes [72].

7.2.8. The System AA'BB'X

Proton spectra of *p*-chloro- and *p*-bromo-fluorobenzene have been studied [60]. They are the AA'BB' parts of an AA'BB'X system. The spectra consist of two apparently identical aa'a''a''' subspectra separated by approximately $(^1/_2)[(J_{AX} + J_{BX}) + 2(D_{AX} + D_{BX})]$. Fig. 22 shows the spectrum of the chloro compound. The lines $1-5$ and $11-15$ belong to one sub-spectrum, and $6-10$

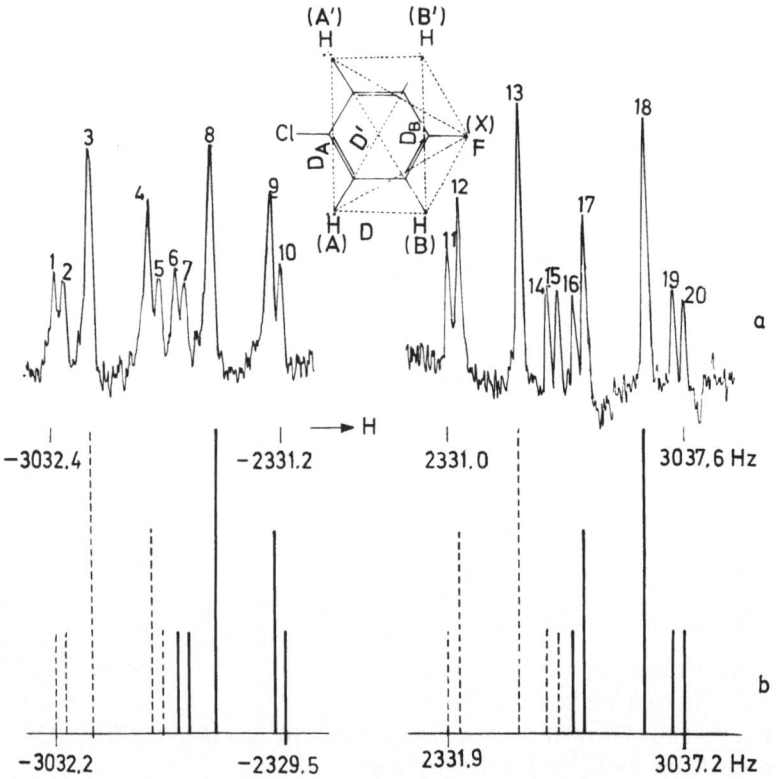

Fig. 22. Observed (a) and calculated (b) proton magnetic resonance spectra of *p*-chloro-fluorobenzene in the nematic phase of 50 % (I) and 50 % (II) [60]. Concentration = 21.6 mole %, temperature = 27° C. Spectrometer frequency = 60 MHz. The two AA'BB' subspectra are indicated in (b)

and $16-20$ to the other. The spectra are 'deceptively simple'. This means that the sub-spectra are degenerate and the effective chemical shift (δ_{eff}) cannot be measured. Conditions for the occurrance of deceptive simplicity in AA'BB'X-spectra are summarised in eqs. (58) to (61) [60] for negligible indirect coupling constants:

$$\{3(D_A - D_B)\delta_{eff}\}/(D + D') < \Delta_{1/2} \tag{58}$$

$$(D_A - D_B)\delta_{eff}/(D - D') < \Delta_{1/2} \tag{59}$$

$$\delta_{eff}^2/2(D - D')^2 < i \tag{60}$$

$$[\delta_{eff} + (^3/_2)(D_A - D_B)]^2/(D + D')^2 < i \tag{61}$$

where $\Delta_{1/2}$ is the line-width, i the minimum observable relative intensity and $\delta_{eff} = \delta_{AB} + (^1/_2)[(J_{AX} - J_{BX}) + 2(D_{AX} - D_{BX})]$.

Conditions (58) to (61) are fulfilled in the spectra of p-substituted fluorobenzenes since $D_A \approx D_B$ and $\delta_{eff} \ll (D + D')$ or $(D - D')$ (symbols are explained in Fig. 22). The spectra do not permit an accurate evaluation of δ_{eff}, δ_{AB}, $(D_{AX} - D_{BX})$, D_A and D_B. Only the parameters reproduced in Table 23 can be obtained.

Table 23. *Direct couplings D, D' and $(D_{AX} + D_{BX})$ obtained from the spectra of p-chloro-and p-bromo-fluorobenzenes*

Compound	Liquid crystal	Temp. (°C)	Concentration (mole %)	D (Hz)	D' (Hz)	$D_{AX} + D_{BX}$ (Hz)
p-chloro-fluorobenzene	50% (I) +50% (II)	27	21.6	-1788 ± 1	-10.2 ± 0.2	-384.0 ± 0.3
p-bromo-fluorobenzene	50% (I) +50% (II)	27	16.1	-1934 ± 1	-7.3 ± 0.2	-393.0 ± 0.3

Deceptive simplicity thus imposes a further practical limitation on the study of molecular geometry from NMR spectra of partially oriented molecules.

7.2.9. The Six Spin System AA'A''A'''A''''A''''' with C_6-Symmetry

Proton and fluorine spectra of benzene and hexafluorobenzene respectively [7, 15, 27, 30, 33, 32] have been studied.

As a consequence of the six-fold symmetry in such molecules, one S-value is sufficient to describe the orientation. There are three different direct couplings D_o, D_m and D_p corresponding to ortho, meta and para nuclei respectively.

The spectrum of benzene has been investigated in the nematic phase of liquid crystals [7, 30, 33] and in cholesteric [15] as well as in lyotropic [27] mesophases. Fig. 23 shows the spectrum in a nematic phase [7]. Group theory [30] and the computer simulation method [33] have been applied in the analysis.

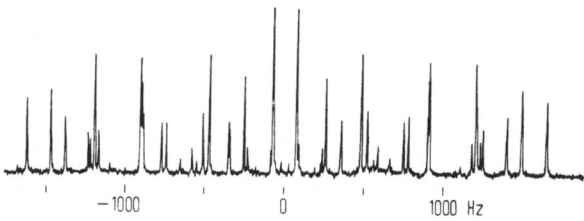

Fig. 23. PMR spectrum of benzene in the nematic phase of 60% (XI) and 40% (XII) [7]. Temperature = 50°C. Spectrometer frequency = 100 MHz. (Reproduced from Angew. Chem. **80**, 99 (1968), copyright (1968). Reprinted by permission of the copyright owner)

Within limits of accuracy, the direct couplings in benzene have been found to agree with regular hexagonal geometry ($D_o : D_m : D_p = 1:0.1925:0.1250$) if the influence of vibrations on the inter-nuclear distances is not considered. This influence has been discussed theoretically [81]. If the vibrational corrections as described in section 4.7 are included and the anharmonic vibrational contribution (linear term in Table 3) is neglected, the following ratios are obtained: $D_o : D_m : D_p = 1:0.1898:0.1225$ with $(r_e)_o = 2.481$ Å, $(r_e)_m$ 4.297 Å and $(r_e)_p = 4.962$ Å. They deviate significantly from the corresponding values determined from NMR [30, 33]. The deviations may be attributed to anharmonic vibrations which cancel the harmonic contributions (quadratic term in Table 3). (The C−H stretching vibration has been shown to possess a significant anharmonicity [84]). Contributions from anisotropic indirect couplings are not considered since they would have to be unfeasibly large.

Direct and indirect couplings have been shown to assume opposite signs in benzene. Since the major indirect proton-proton couplings are known to be positive, the direct ones must be negative. This indicates that the molecule orients preferentially with its plane in the direction of the magnetic field. In cholesteric and lyotropic mesophases, on the other hand, the molecule has been shown to orient preferentially with its plane perpendicular to the magnetic field.

The ^{19}F spectrum of hexafluorobenzene has been investigated in the nematic phase of (III) [32]. D_o, D_m and D_p are not found in the ratio of the inverse cubes of the inter-nuclear distances expected for a rigid, regular hexagonal geometry. This has been attributed partly to the out of plane motions of fluorines and partly to the contributions from anisotropy of the indirect couplings.

For direct couplings the signs found are identical to that of the ortho fluorine-fluorine indirect coupling. If it is assumed that the molecule orients preferentially with its plane in the direction of the magnetic field, as do other aromatics, the sign of the ortho F−F indirect coupling must be negative.

7.2.10. The System of the $A_3A_3'(A_3X_3)$ Type with C_3-Symmetry

Spectra of molecules containing two rotating CX_3 groups (CH_3 or CF_3 for example) belong to this class. They are defined by two direct and one indirect couplings: (i) D, the direct coupling between the nuclei of the two CX_3 groups defined by:

$$D = -\frac{h\gamma_i\gamma_j}{4\pi^2}\left\langle\frac{S}{R^3}\right\rangle \tag{62}$$

with $S = (^1/_2)\,S_{C_3}\,(3\cos^2\alpha - 1)$ where α is the variable angle between the C_3-symmetry axis and the inter-nuclear distance vector (R); (ii) D', the direct coupling between the nuclei within the same CX_3-group, given by:

$$D' = -\frac{h\gamma_i\gamma_j}{4\pi^2 r^3}\cdot(^1/_2)\,S_{C_3} \tag{63}$$

and (iii) J, the indirect coupling between the nuclei of the two CX_3 groups.

Indirect coupling within the same group does not affect the spectrum.

Frequencies and intensities of the allowed transitions are analytical functions of the parameters and are reproduced in Table 24 [45].

Table 24. *Frequencies and intensities of the allowed transitions in an A_3A_3' system [45] derived using group theory for non-rigid molecules [85]*
Frequencies are relative to $\nu_{A_3} = 0$. One half of the symmetrical spectrum is given

No.	Transition	Frequency	Relative intensity
1	$(Ag)_2 \leftarrow (Ag)_3$	$-3D' - (^9/_2)D$	6
2	$1(Ag)_1 \leftarrow (Ag)_2$	$-(^3/_2)D' - (^1/_2)D - (^5/_2)J + ^1/_2\,D_1$	$(2a_1 + \sqrt{6}a_2)^2$
3	$2(Ag)_1 \leftarrow (Ag)_2$	$-^3/_2\,D' - (^1/_2)D - (^5/_2)J - (^1/_2)D_1$	$(-2a_2 + \sqrt{6}a_1)^2$
4	$1(Ag)_0 \leftarrow 1(Ag)_1$	$(^3/_2)D' - 3D + (^1/_2)(3D_0 - D_1)$	$(\sqrt{3}a_1b_1 + \sqrt{3}a_1b_2 + 2\sqrt{2}a_2b_2)^2$
5	$2(Ag)_0 \leftarrow 1(Ag)_1$	$(^3/_2)D' - 3D - ^1/_2(3D_0 + D_1)$	$(-\sqrt{3}a_1b_2 + \sqrt{3}a_1b_1 + 2\sqrt{2}a_2b_1)^2$
6	$1(Ag)_0 \leftarrow 2(Ag)_1$	$(^3/_2)D' - 3D + ^1/_2(3D_0 + D_1)$	$(-\sqrt{3}a_2b_1 - \sqrt{3}a_2b_2 + 2\sqrt{2}a_1b_2)^2$
7	$2(Ag)_0 \leftarrow 2(Ag)_1$	$(^3/_2)D' - 3D - (^1/_2)(3D_0 - D_1)$	$(\sqrt{3}a_2b_2 - \sqrt{3}a_2b_1 + 2\sqrt{2}a_1b_1)^2$
8	$(Au)_1 \leftarrow (Au)_2$	$-(^9/_2)D$	4
9	$1(Au)_0 \leftarrow (Au)_1$	$-(^3/_2)J + (^3/_2)E$	$3(c_1 + c_2)^2$
10	$2(Au)_0 \leftarrow (Au)_1$	$-(^3/_2)J - (^3/_2)E$	$3(-c_1 + c_2)^2$
11	$1(G_1)_1 \leftarrow (G_1)_2$	$-(^3/_2)D' - 2D - J + (^1/_2)F$	$4(d_1 + \sqrt{3}\,d_2)^2$
12	$2(G_1)_1 \leftarrow (G_1)_2$	$-(^3/_2)D' - 2D - J - (^1/_2)F$	$4(-d_2 + \sqrt{3}\,d_1)^2$
13	$(G_{1+})_0 \leftarrow 1(G_1)_1$	$-(^3/_2)D' - D + J - (^1/_2)F$	$2(\sqrt{3}\,d_1 + 3d_2)^2$
14	$(G_{1-})_0 \leftarrow 1(G_1)_1$	$-(^3/_2)D' + D - J - ^1/_2F$	$2(\sqrt{3}\,d_1 - d_2)^2$
15	$(G_{1+})_0 \leftarrow 2(G_1)_1$	$-(^3/_2)D' - D + J + ^1/_2F$	$2(-\sqrt{3}\,d_2 + 3d_1)^2$
16	$(G_{1-})_0 \leftarrow 2(G_1)_1$	$-(^3/_2)D' + D - J + ^1/_2F$	$2(\sqrt{3}\,d_2 + d_1)^2$
17	$(G_{2g})_0 \leftarrow (G_{2g})_1$	$-(^3/_2)D$	8

Where:

$D_1 = [(3D' - 2D - J)^2 + 24(D - J)^2]^{1/2}$; $E = [(2D' - 2D)^2 + (D - J)^2]^{1/2}$

$D_0 = [(2D' - (\frac{2}{3})D - (\frac{4}{3})J)^2 + (D - J)^2]^{1/2}$; $F = [(3D' - 2D - J)^2 + 3(D - J)^2]^{1/2}$

$$\frac{a_1}{a_2} = -\frac{[3D'-2D-J+D_1]}{2\sqrt{6}(D-J)}; \ a_1^2 + a_2^2 = 1; \ \frac{b_1}{b_2} = -\frac{[2D'-(\frac{2}{3})D-(\frac{4}{3})J+D_0]}{(D-J)}, \ b_1^2 + b_2^2 = 1$$

$$\frac{c_1}{c_2} = -\frac{[2D'-2D+E]}{(D-J)}, \ c_1^2 + c_2^2 = 1; \ \frac{d_1}{d_2} = -\frac{[3D'-2D-J+F]}{\sqrt{3}(D-J)}, \ d_1^2 + d_2^2 = 1$$

Spectra of 2,4-hexadiyne [45], dimethyl acetylene [86] and perfluorodimethyl acetylene [86] have been studied. The spectrum of dimethyl acetylene (Fig. 24) has been investigated at 100 MHz with and without spinning and at 220 MHz with spinning.

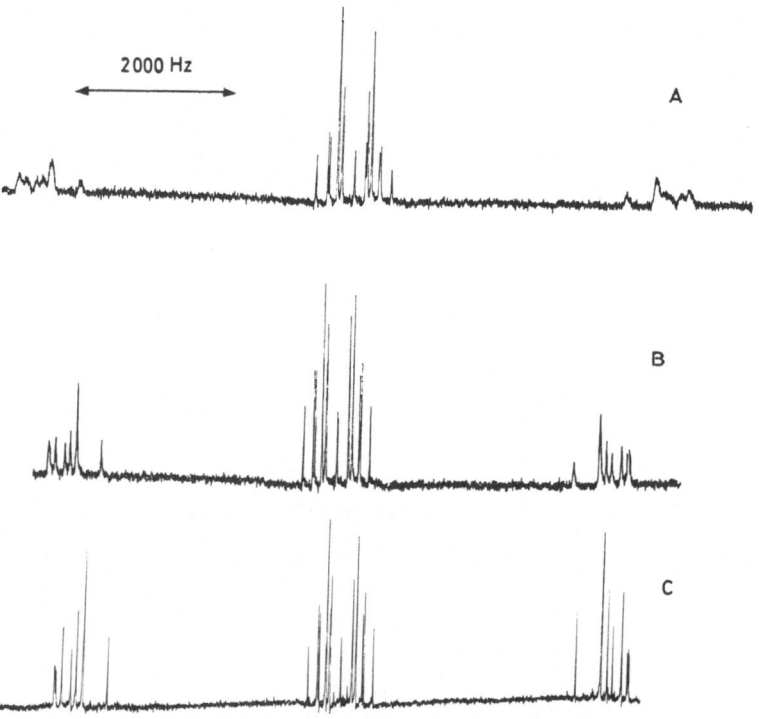

Fig. 24. PMR spectra of dimethyl acetylene in the nematic phase of (III) [86] at 80°C. (A) 100 MHz spectrum without spinning, (B) 100 MHz spectrum, spinning speed = 6 Hz, (C) 220 MHz spectrum, spinning speed = 100–150 Hz. (Reprinted from Mol. Phys. 15, 285 (1968), copyright (1968). Reproduced by permission of the copyright owner)

The Varian 220 MHz spectrometer has a superconducting magnet with the field direction parallel to the spinning axis. Spinning of the sample does not destroy the orientation. Drastic improvement of line-widths and signal to noise ratio is achieved (Fig. 24).

The spectrum of 2,4-hexadiyne has been analysed on the basis of the relations given in Table 24. The sign of J has been found to be positive in this case.

Spectra of dimethyl acetylene and perfluorodimethyl acetylene have been analysed with the help of the modified FREQUINT computer programme.

Because of a C_3-symmetry axis only one S-value is required to describe the orientation. D/D' is independent of the orientation parameter and can be used for the determination of distance ratios (eqs. (64), (65) and (66)) [45]. As a consequence of the internal rotation about the $C-C$ bond, D is averaged over the relative rotational angles of the CX_3-groups.

$$(D/D') = -\frac{3\sqrt{3}}{\pi} \int_0^{2\pi} \frac{(\beta - 1 + \cos\varphi)\,d\varphi}{(\beta + 2 - 2\cos\varphi)^{5/2}} \qquad (64)$$

where $\beta = (3\,\xi_0^2/r^2) = (\xi_0^2/\rho_0^2)$ such that ρ is the distance from the X-nuclei to the C_3-axis of symmetry and ξ_0 is the separation between the two parallel planes containing the three nuclei of the same CX_3 group. The angle φ is the relative rotational angle between the CX_3 groups and r is the distance between the X-nuclei within the CX_3 group.

The distance R between the protons of the two methyl groups is related to ξ_0 and ρ by (65) and (66):

$$R^2 = \xi_0^2 + 2\rho^2(1 - \cos\varphi) \qquad (65)$$

$$R^2 = \xi_0^2/\cos^2\alpha \qquad (66)$$

The values of β may be obtained by numerical computation of the integral (64). In the molecules studied, it is then possible to obtain the HCH (FCF) angle from known values of the CH and CC bond lengths. In Table 25, the values obtained for the three compounds are compared with electron diffraction and microwave data.

Table 25. *D/D' and HCH(FCF) angles for 2,4-hexadiyne, dimethyl acetylene and perfluoro-dimethyl acetylene*

Compound	(D/D')	HCH(FCF) angle estimated from NMR	HCH(FCF) angle obtained from other sources	Reference
2,4-hexadiyne	-0.0266 ± 0.0002	$110°10'$ [a]	$108°25'$ (value for propyne) [87]	[45]
dimethyl acetylene	-0.0801 ± 0.0009	$109.1°$ [b]	$108.3°$ [88, 89]	[86]
perfluorodimethyl-acetylene	-0.1235 ± 0.0005	$110.7°$ [c]	$107.5°$ [90, 91]	[86]

  [a] Assumed CH and $CC\equiv C-C\equiv C-C$ bond lengths are $1.1124\,\text{Å}$ and $6.693\,\text{Å}$ respectively.

  [b] Assumed CH and $CC\equiv CC$ bond lengths are $1.1124\,\text{Å}$ and $4.1227\,\text{Å}$ respectively.

  [c] Assumed CF and $CC\equiv CC$ bond lengths are $1.34\,\text{Å}$ and $4.150\,\text{Å}$ respectively.

Agreement is satisfactory for dimethyl acetylene only. 2,4-Hexadiyne has been compared with propyne. The discrepancy observed is attributed to a possible increase of the angle HCH caused by the additional triple bond in the former. In the perfluoro compound, the discrepancy may be due to the omission of the anisotropic contributions to the indirect $F-F$ couplings.

The spectra of ethane [76, 141], 1,1,1-trifluoroethane [76, 141] and mercury dimethyl [136] have recently been investigated.

Only the staggered configuration of ethane conforms with the observed spectrum [76].

In 1,1,1-trifluoroethane, an agreement between NMR and microwave values for r_{FF}/r_{HH} (1.192 and 1.211 respectively) indicates a negligible anisotropy in the $F-F$ indirect coupling [76].

In mercurydimethyl, the NMR data have not allowed discrimination between free rotation and the staggered or eclipsed conformations of the methyl group, the differences in D/D' and β being too small. D/D' has been found to be 0.07875 ± 0.0001 and the absolute sign of J positive. Additional sub-spectra are observed due to negative direct and indirect couplings of ^1H with ^{199}Hg in natural abundance (approximately 17%). No anisotropic contributions to the latter have been observed.

7.2.11. The AA'A''A'''A''''A''''' Type with D_{3h}-Symmetry

Cyclopropane has a plane of symmetry containing the three carbon atoms and a three-fold symmetry axis. It has three independent direct coupling constants. One S-value completely describes the orientation. The spectrum of the molecule has been investigated extensively [10, 34, 92]. Fig. 25 shows the lower field half of the spectrum which is symmetric about the centre and extends over 2040 Hz. Analysis has been performed by the computer simulation method.

Spectra of the protons attached to ^{13}C nuclei in natural abundance have also been recorded by means of a time averaging computer.

Values of the direct couplings derived are reproduced in Table 26.

Table 26. *Direct coupling values obtained from the spectra of cyclopropane* [92]. *Numbering is explained in Fig. 25*

D_{ij}	Value (Hz)
D_{12}	-97.26
D_{14}	$+487.00$
D_{15}	$+2.18$
$D_{^{13}C_{(1)}H_{(1)}}$	$+325.89$
$D_{^{13}C_{(1)}H_{(3)}}$	-16.98

The indirect coupling between ^{13}C and a directly bonded proton has been found to be positive. The influence of J_{14} and the indirect coupling of ^{13}C to a proton bonded to a neighbouring carbon atom has been shown to be too small for evaluation.

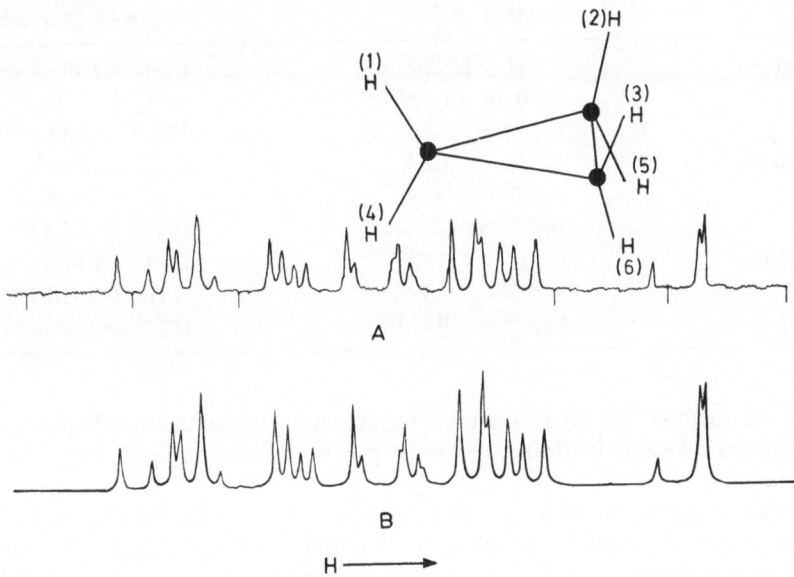

Fig. 25. Low field half of the spectrum of cyclopropane in the nematic phase of (III) [92]. (A) Experimental (B) computer simulation. Concentration = 25 mole %. Temperature = 75° C. Spectrometer frequency = 60 MHz. Frequency markers shown are spaced at 106.25 Hz. (Reprinted from J. Chem. Phys. **47**, 1480 (1967), copyright (1967). Reproduced by permission of the copyright owner)

Different pairs of direct couplings have been used in conjunction with electron diffraction data [93] for $C-C$ and $C-H$ bond lengths in the study of molecular geometry (Table 27).

The influence of molecular vibration on the spectrum of cyclopropane has been discussed in terms of D_{ij}^{pseudo} (eq. 2). In a first approximation some of the D_{ij}^{pseudo} are set equal to zero. The others together with the equilibrium geometry are then adjusted until a fit is obtained. The process is repeated setting other D_{ij}^{pseudo} values to zero. The results are included in Table 27. The errors are indicative of the uncertainties in the determination of molecular geometry when the influence of vibrations is neglected. A value of the pseudo dipolar coupling equal to $-11.8\,Hz$ has also been obtained for the interaction between the two methylene groups in allene [76].

Table 27 indicates a reduction of the bond length $C-C$ relative to $C-H$. This has been assigned to large methylene 'rocking' motions which shorten the distance between the protons on the same side of the ring, i.e. the distance from which the $C-C$ bond lengths are computed under the assumption of D_{3h}-symmetry.

Table 27. *Structure of cyclopropane*

Direct couplings fitted	Information derived	
	C−C bond length = 1.510 Å assumed [93]	C−H bond length equal to 1.089 Å assumed [93]
D_{12} and D_{15}	C−H = 1.122 Å	C−C = 1.465 Å
	< HCH = 113.6°	< HCH = 113.6°
	$D_{14}^{pseudo} = -13.55$ Hz	$D_{14}^{pseudo} = -13.55$ Hz
D_{14} and D_{15}	C−H = 1.107 Å	C−C = 1.485 Å
	< HCH = 114.4°	< HCH = 114.4°
	$D_{12}^{pseudo} = -2.4$ Hz	$D_{12}^{pseudo} = -2.4$ Hz
D_{12} and D_{14}	C−H = 1.123 Å	C−C = 1.464 Å
	< HCH = 114.4°	< HCH = 114.4°
	$D_{15}^{pseudo} = -0.61$ Hz	$D_{15}^{pseudo} = -0.61$ Hz

Of the various combinations used for the study of geometry (Table 27), the last is preferred since it utilizes the most precise direct couplings of the largest magnitude.

In this category, the spectrum of s-trioxane has also been investigated [127]. The molecule may exist in the 'chair' or the 'boat' conformations. Whereas the former has a threefold axis of symmetry (and hence only one S-value defines orientation), the latter has only a plane of symmetry (and, therefore, three orientation parameters are necessary). Analysis of the spectrum shows that the molecule is predominantly in the 'chair' form as also indicated by the dipole moment [128], electron diffraction [129, 130] and microwave [131] studies. Each of the methylene groups is found to be tilted by 2°50′ in the direction which increases the distance between the axial protons. This means that the OCO plane does not bisect the HCH angle. $J_{^{13}C-H}$ has been found to be positive in this molecule.

7.2.12. The System AA′A″XX′X″ with D_{3h}-Symmetry

Proton and fluorine magnetic resonance spectra of 1,3,5-trifluorobenzene have been reported [94, 95]. One half of each symmetrical spectrum is shown in Fig. 26. Parameters obtained from the analysis are reproduced in Table 28.

Table 28. *Parameters obtained from ^1H and ^{19}F spectra of 1,3,5-trifluoro-benzene [94]*

Nuclear pair	D_{ij} (Hz)	J_{ij} (Hz)	D_{ij}^{pseudo} (Hz)
F−F	−76.0	5.8	0.5
H−H	−112.15	2.3	1.0
(H−F)$_{ortho}$	−474.05	9.0	−
(H−F)$_{para}$	−59.8	−1.7	−0.5

It has been found that the indirect couplings (J_{ij}) do not deviate from those measured in isotropic media [96]. In order to obtain the best overall fit for ^{1}H and ^{19}F spectra, pseudo dipolar couplings (Table 28) had to be introduced.

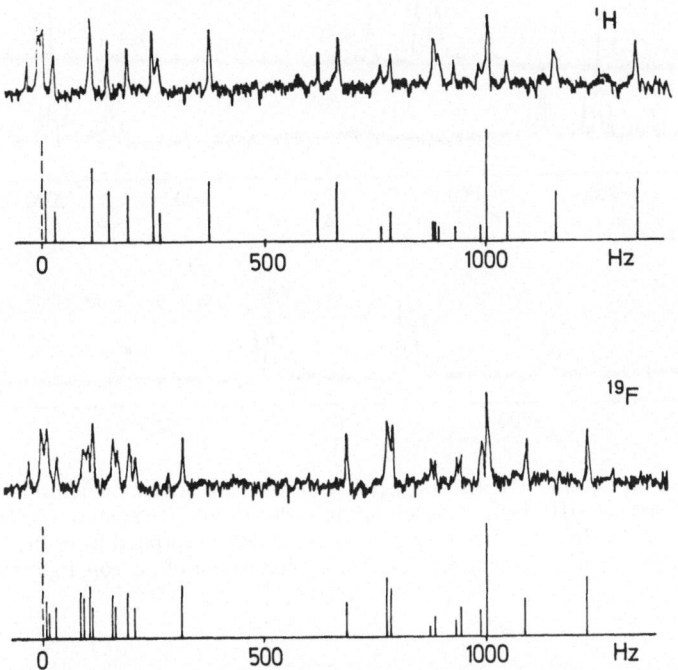

Fig. 26. Proton and fluorine magnetic resonance spectra of 1,3,5-trifluorobenzene in the nematic phase of a mixture of 3.5% (XIV), 34.8% (XV) and 61.7% (III) [94]. Concentration = 30 mole%. Temperature = 62° C. Spectrometer frequency = 56.4 MHz. The spectra are symmetrical about the centre and only one half is shown in each case. (Reproduced by permission of the National Research Council of Canada from the Can. J. Chem. 46, 2783 (1968))

Known bond lengths, (CC = 1.390 Å and CH = 1.080 Å), combined with the data of Table 28 provide a C−F bond length of 1.296 Å. This agrees with the electron diffraction value [97] for meta and para difluorobenzene but is smaller than that derived from microwave data [98] on monofluorobenzene (1.33 Å).

7.2.13. The System of the Type AA′A″A‴XX′ with D_{2h}-Symmetry

Para difluoro- and symmetrical tetrafluoro-benzene have D_{2h}-symmetry. Their spectra are shown in Figs. 27 and 28 respectively [48].

The proton spectrum of p-difluorobenzene is deceptively simple [60]. It consists of three aa′a″a‴ type sub-spectra with relative intensity 1:2:1 and spacing $(J_{HF}^o + J_{HF}^m + 2(D_{HF}^o + D_{HF}^m))$. The direct couplings D_{HF}^o and D_{HF}^m cannot be determined individually. From equations similar to (42−44) and (47) or those reported in reference [48] the ratios r_{HH}^m/r_{HH}^o and r_{FF}^p/r_{HH}^o are found to be 1.6975 and 2.1335 respectively.

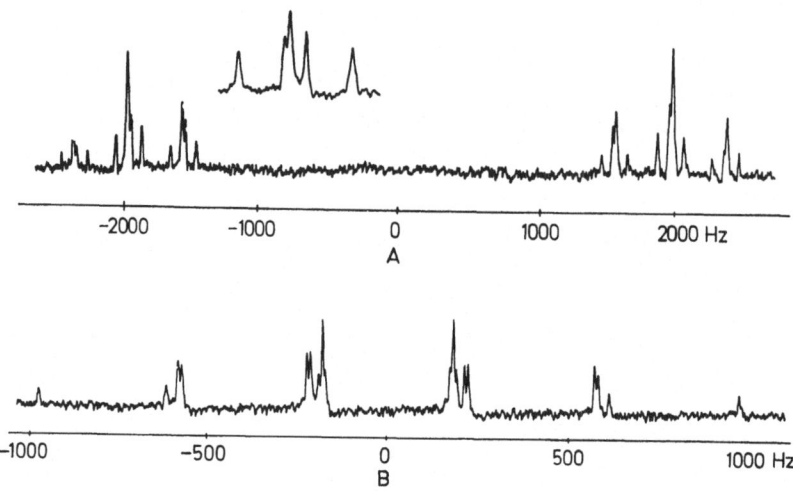

Fig. 27. Proton (A) and ^{19}F (B) magnetic resonance spectra of p-difluorobenzene in the nematic phase of (III) [48]. Concentration = 40 mole%. Temperature = 80°C for (A) and 75°C for (B). Spectrometer frequency = 56.4 MHz. (Reprinted from Rec. Trav. Chim. 87, 417 (1968), copyright (1968). Reproduced by permission of the copyright owner and the authors)

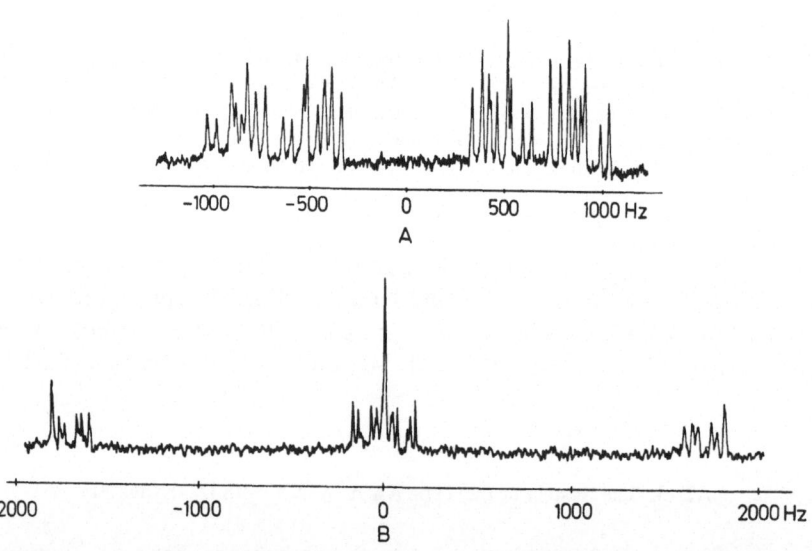

Fig. 28. ^{19}F (A) and ^{1}H (B) magnetic resonance spectra of symmetrical tetrafluorobenzene in the nematic phase of (III) [48]. Concentration = 37 mole % for (A) and 20 mole % for (B). Temperature = 68°C for (A) and 70°C for (B). Spectrometer frequency = 56.4 MHz. (Reprinted from Rec. Trav. Chim. 87, 417 (1968), copyright (1968). Reproduced by permission of the copyright owner and the authors)

The value of (r_{HH}^m/r_{HH}^o) deviates significantly from 1.7321 which is expected for a regular hexagonal symmetry.

The parameters obtained for symmetrical tetrafluorobenzene are presented in Table 29.

Table 29. *Values of the parameters obtained from the spectrum of an 18 mole % solution of symmetrical tetrafluorobenzene at 70° C in the nematic phase of (III). Values of the indirect couplings are:*
$J_{FF}^o = -17.8$, $J_{FF}^m = 4.0$, $J_{FF}^p = 11.1$,
$J_{HF}^o = 10.0$, $J_{HF}^m = 8.0$ Hz

Parameter	Value (Hz)
D_{FF}^o	23.5 ± 0.5
D_{FF}^m	-181.0 ± 0.5
D_{FF}^p	-93 ± 1.0
D_{HF}^o	-909 ± 1.5
D_{HF}^m	59 ± 1.5
D_{HH}	2.75 ± 0.5
$(D_{FF}^{ind})_{ortho}$	9 ± 4
$(J_{xx} - J_{zz})_{FF}^{ortho}$	138 ± 54
$(2J_{xx} + J_{yy})_{FF}^{ortho}$	85 ± 57

y axis is the H−H axis, z axis is perpendicular to the ring plane

Again the direct couplings deviate from those of a regular hexagonal geometry; this is attributed to an anisotropic contribution in the indirect coupling $(D_{FF}^{ind})_{ortho}$. (D_{FF}^o) may be computed from the observed D_{HH} on the basis of a regular hexagonal geometry ($r_{CC} = 1.935 \pm 0.003$ Å, $r_{CF} = 1.33 \pm 0.04$ Å, $r_{CH} = 1.084 \pm 0.007$ Å). It is compared with the experimental value reported in Table 29. The difference $(D_{FF}^{ind})_{ortho}$ is $(9 \pm 4$ Hz$)$. If $(D_{HF}^{ind})_{ortho}$ and $(D_{HF}^{ind})_{meta}$ are assumed to be zero, relations between the components of the ortho F−F indirect coupling may be deduced (Table 29).

7.2.14. The System AA'BB'CC' with C_{2v}-Symmetry

An example of the system AA'BB'CC' with C_{2v}-symmetry is the proton spectrum of bicyclobutane. The molecule has seven different types of direct couplings, and two parameters are required for a definition of its orientation. The spectrum [10, 99] is shown in Fig. 29. INDOR experiments had to be performed for line-assignment in this complex molecule. In addition, a time averaging computer has been used to record the relatively weak spectra of molecules containing ^{13}C in natural abundance. Such spectra permit the location of the protons relative to the carbon atoms. Table 30 reproduces the parameters derived.

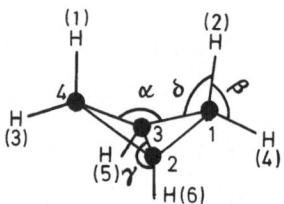

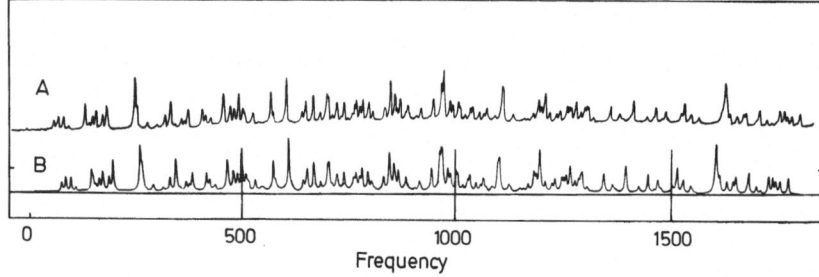

Frequency

Fig. 29. PMR spectrum of bicyclobutane in a liquid crystal solvent [*10*]. (A) Theoretical (B) Experimental. (Reproduced from Science **162**, 1337 (1968), copyright (1968), by the American Association for the Advancement of Science. Reprinted by permission of the copyright owner and the authors)

Table 30. *Direct and indirect coupling constants for bicyclobutane in a liquid crystal solvent* [*10, 99*].
(Symbols are explained in Fig. 29).
$(v_1 - v_3) = 29.8\,Hz,\ (v_5 - v_3) = -29.4\,Hz$

Parameter	Proton pair						
	1,2	1,3	1,4	1,5	3,4	3,5	5,6
Indirect coupling (Hz)	2.0	0.0	0.6	−0.1	4.9	2.8	7.5
Direct coupling (Hz)	−383.3	142.9	−76.9	48.7	−74.0	−113.7	123.55

Geometric information obtained in bicyclobutane is summarised in Table 31 together with that determined from electron diffraction data.

Table 31. *Geometry-information in bicyclobutane* [10]

Parameter	Source	
	NMR	Electron diffraction
$<\alpha$	$120.2 \pm 2°$	$122.8°$
$<\beta$	$110.2 \pm 1°$	$116.0°$
$<\gamma$	$128.0 \pm 2°$	$125.5°$
$<\delta$	$126.3 \pm 1°$	$\approx 122°$
$C_{(1)} - C_{(2)}$	1.507 Å^a	1.507 Å
$C_{(2)} - C_{(3)}$	$1.507 \pm 0.07 \text{ Å}$	1.502 Å
$C_{(1)} - H_{(2)}$	$1.167 \pm 0.02 \text{ Å}$	$\approx 1.106 \text{ Å}$
$C_{(1)} - H_{(4)}$	$1.194 \pm 0.02 \text{ Å}$	$\approx 1.106 \text{ Å}$
$C_{(2)} - H_{(6)}$	$1.142 \pm 0.02 \text{ Å}$	1.108 Å

[a] Electron diffraction value (assumed).

7.2.15. The System AA′BB′CX with C_{2v}-Symmetry

Fluorobenzene and monodeuterobenzene which are planar and have C_{2v}-symmetry have been studied [31, 38].

The ^1H and ^{19}F magnetic resonance spectra of fluorobenzene have been analysed by the computer simulation method under the following assumptions: i) the aromatic ring is a regular hexagon with CC and CH bond lengths equal to 1.39 and 1.08 Å respectively, ii) values of the proton chemical shift used have been taken from experiments in isotropic media and iii) D_{ij}^{ind} is zero. The assumptions are justified for the experimental accuracy reported [31]. However under these conditions, it is not possible to derive precise information about the molecular geometry. Moreover, some of the D-values have been shown to be insensitive to small structural changes [31].

Table 32. *Direct couplings, chemical shifts,* S_{22}/S_{11}, r_{16}/r_{12} *and* $(r_{CH} - r_{CD})/r_{12}$ *in monodeutero-benzene* [38]

Parameter	Value	Parameter	Value
$D_{12} = D_{45}$	$-393.7 \pm 0.7 \text{ Hz}$	D_{15}	$-76.3 \pm 2.1 \text{ Hz}$
$D_{23} = D_{34}$	$-396.1 \pm 0.4 \text{ Hz}$	D_{36}	$-6.9 \pm 0.3 \text{ Hz}$
$D_{16} = D_{56}$	$-61.5 \pm 0.7 \text{ Hz}$	$\nu_2 - \nu_1$	$0.1 \pm 0.3 \text{ Hz}$
$D_{13} = D_{35}$	$-76.4 \pm 0.4 \text{ Hz}$	$\nu_3 - \nu_1$	$0.4 \pm 0.4 \text{ Hz}$
$D_{26} = D_{46}$	$-11.6 \pm 0.3 \text{ Hz}$	S_{22}/S_{11}	0.992 ± 0.004
$D_{14} = D_{25}$	$-50.3 \pm 0.5 \text{ Hz}$	r_{16}/r_{12}	0.996 ± 0.004
D_{24}	$-75.5 \pm 0.5 \text{ Hz}$	$(r_{CH} - r_{CD})/r_{12}$	0.008 ± 0.008

The proton spectrum of monodeuterobenzene (Fig. 30) [*38*] is a superposition of three 5-spin sub-spectra due to the weak coupling with the deuteron ($I = 1$). If the deuteron substitution does not produce a detectable chemical shift between the various protons, the effective chemical shifts are $(2D_{HD} + J_{HD})$, 0 and $-(2D_{HD} + J_{HD})$ for the possible deuteron spin states 1,0 and -1 respectively.

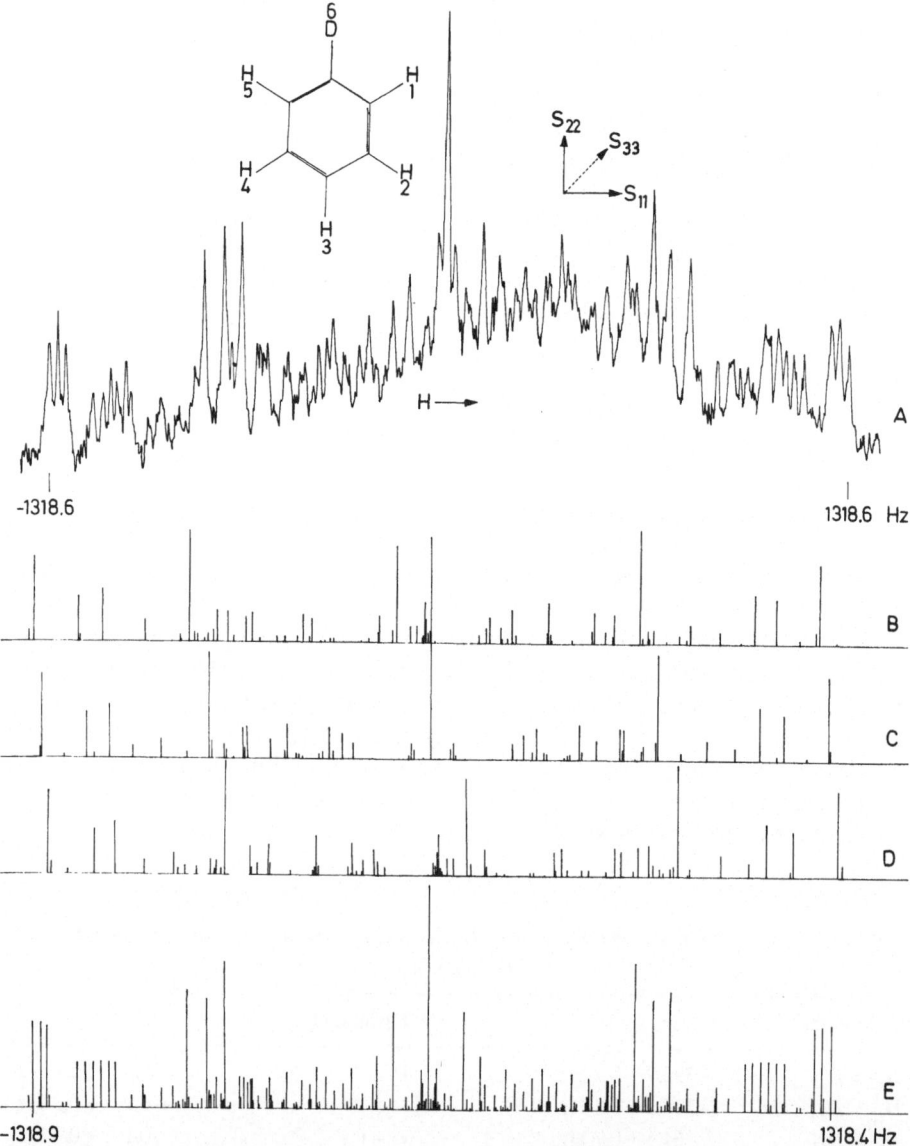

Fig. 30. Proton magnetic resonance spectra of a 24 mole% solution of monodeuterobenzene at 27°C in the nematic phase of (I) + (II). (A) observed spectrum; (B) the 5 spin sub-spectrum corresponding to $m(D) = 1$; (C) the five spin sub-spectrum due to $m(D) = 0$; (D) the five spin sub-spectrum involving the state $m(D) = -1$ and (E) the calculated spectrum (superposition of B, C and D [*38*].

The spectrum has been analysed iteratively with the deuteron treated as two spin $^1/_2$ particles. An intensity correction had to be applied for the spin zero state. Calculated spectra are included in Fig. 30.

Table 32 summarises the various parameters.

Table 32 shows that D_{12} and D_{23} are not equal. Both are smaller than $(\gamma_H/\gamma_D) D_{16}$ (-400.6 ± 4.5 Hz) in which (γ_H/γ_D) has been taken to be 6.5144. These findings have been attributed to deviations from a regular hexagonal geometry of the molecule. As a consequence of the effective C_{2v}-symmetry, two orientation parameters S_{11} and S_{22} (Fig. 30) are essential in the specification of molecular orientation. They are related to D_{12}, D_{23} and the ortho proton-proton distance (r_{HH}^{ortho}) by eqs. (67) and (68):

$$S_{11} = -\tfrac{4}{3} \left[\frac{4\pi^2 (r_{HH}^{ortho})^3}{h \gamma_H^2} \right] (D_{23} - (^1/_4) D_{12}) \tag{67}$$

$$S_{22} = - \left[\frac{4\pi^2 (r_{HH}^{ortho})^3}{h \gamma_H^2} \right] D_{12} . \tag{68}$$

S_{11} and S_{22} can be determined separately but with low precision from the known r_{HH}^{ortho} value in benzene (2.481 ± 0.005 Å [100]). However, the ratio S_{22}/S_{11} given by:

$$(S_{22}/S_{11}) = (^3/_4) \cdot \frac{D_{12}}{(D_{23} - (^1/_4) D_{12})}$$

is independent of r_{HH}^{ortho} and can therefore be obtained precisely (Table 32). The value is significantly different from unity. This may mean a reduced orientation of the C_2-symmetry axis as a result of deuteron substitution. Provided that neither the geometry of the carbon ring nor that of the remaining five protons is changed, the relations between the various H $-$ D distance ratios and the direct couplings are given by eqs. (70) and (71):

$$(\gamma_H/\gamma_D) D_{i6} (r_{i6}/r_{12})^5 = \{(r_{i6}/r_{12})^2 - (^3/_4)\} \cdot D_{12} + (D_{23} - (^1/_4) D_{12}) \tag{70}$$

where $i = 1$ or 2, and

$$((\gamma_H/\gamma_D) D_{36} (r_{36}/r_{12})^3 = D_{12} . \tag{71}$$

Since D_{26} and D_{36} (Table 32) involve large uncertainties, (r_{26}/r_{12}) and (r_{36}/r_{12}) cannot be computed from eqs. (70) and (71) to a reasonable accuracy. However, the value of (r_{16}/r_{12}) can be obtained precisely (Table 33). (r_{16}/r_{12}) is related to $(r_{CH} - r_{CD})/r_{12}$ as expressed in formula (72):

$$\frac{(r_{CH} - r_{CD})}{r_{12}} = \left\{ (^1/_2) - \left[\left(\frac{r_{16}}{r_{12}} \right)^2 - (^3/_4) \right]^{1/2} \right\} . \tag{72}$$

The results given in Table 32 indicate a slight reduction in the average value of the C $-$ D bond length compared to that of C $-$ H. Similar results have been obtained by several other methods [100 $-$ 102].

7.2.16. The System A_3B_2X

Ethyl fluoride has an effective plane of symmetry if there is a free rotation about the $C-C$ bond. Only average values of the direct couplings between nuclei in different groups can be obtained in this case. Three parameters are needed to describe the orientation of the molecule which has five different direct couplings. The proton and fluorine spectra have been investigated in the nematic phase of a mixture of 40% (XI) and 60% (XII) [103]. Values of the parameters obtained are given in Table 33.

Table 33. *Parameters obtained from the proton and the fluorine spectra of ethyl-fluoride* [103]. *(Values of the indirect couplings used:*
$J_{F-CH_2} = 47.5\,Hz$, $J_{F-CH_3} = 25.7\,Hz$, $J_{CH_2-CH_3} = 6.9\,Hz)$

$$\begin{array}{ccc}
(3)\,H\diagdown & & \diagup H\,(6) \\
(2)\,H\!-\!C\!-\!C\!-\!H\,(5) \\
(1)\,F\diagup & & \diagdown H\,(4)
\end{array}$$

Parameter	Value (Hz)	
	From ^1H spectrum	From ^{19}F spectrum
$D_{12} = D_{13}$	271.5 ± 0.5	283.3 ± 0.2
$D_{14} = D_{15} = D_{16}$	-225.2 ± 0.6	-235.0 ± 0.2
D_{23}	799.5 ± 3	843.0 ± 4.5
$D_{24} = D_{25} = D_{26} = D_{34} = D_{35} = D_{36}$	-213.8 ± 0.6	-225.3 ± 0.8
$D_{45} = D_{46} = D_{56}$	575.0 ± 2	606.5 ± 2.5
$\nu_{CH_3} - \nu_{CH_2}$	313.3 ± 1.0	313.3 (assumed)

Contradictions between the values obtained from the ^1H and ^{19}F spectra have been attributed to differences in the temperature at which the spectra were studied [103].

Geometric information derived from the NMR data in ethylfluoride has been shown to be more consistent with the microwave values reported in references [104] and [105] than in [106].

7.2.17. The 8-Spin System with D_{2d}-Symmetry

The proton spectrum of cyclobutane which belongs to this category is presented in Fig. 31 [10, 34].

Calculated spectra for the various structures of cyclobutane are also shown indicating that a satisfactory fit can be obtained only for two rapidly inter-converting non-planar, bent conformers with D_{2d}-symmetry. The degree of bending has been defined by the dihedral angle between two planes each containing three carbon atoms. Structural information derived is compared with the electron diffraction data [107] in Table 34.

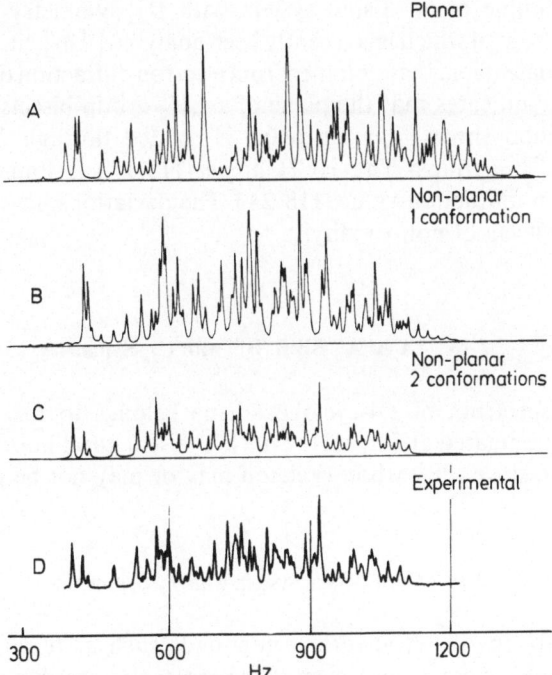

Fig. 31. Proton magnetic resonance spectra of cyclobutane in a nematic solvent [10]. (A), (B) and (C) are the calculated spectra for the molecular structures indicated. (D) is the experimental spectrum. Only half of the symmetrical spectra is shown. (Reproduced from Science **162**, 1337 (1968). Copyright (1968) by the American Association for the Advancement of Science. Reprinted by permission of the copyright owner and the authors)

Table 34. *Structure of cyclobutane* [10, 34]

Parameter	Value obtained from NMR data	Value obtained from electron diffraction data [107]
C−C bond length	1.548 Å (assumed)	1.548 ± 0.003 Å
C−H bond length	1.171 ± 0.02 Å	1.092 ± 0.010 Å
< HCH	108.5 ± 2°	110°
Dihedral angle	35 ± 2° (≈ 28°)[b]	35°
methylene tilt[a] angle	2.5 ± 2°	0° (assumed)

[a] The angle between the bisector of < HCH and the plane through the nearest carbon atoms. The tilt-direction corresponds to an increase in the distance between the axial protons on the same side of the ring.

[b] Recent results [10].

Values of 0.85 and -1.85 Hz have been found for D_{cis}^{pseudo} and D_{trans}^{pseudo} (across the ring) in cyclobutane. As in cyclopropane [92], they have been attributed to large methylene rocking motions.

Another example of an 8-spin system with D_{2d}-symmetry is spiropentane, the proton spectrum of which has recently been analysed [126]. Structural information derived deviates from that obtained from electron diffraction data. Whereas the NMR spectrum indicates that the plane of a CH_2 group bisects the CCC angle, electron diffraction shows that the plane is rotated through 2.3° towards the neighbouring CH_2 groups. The HCH angle (115°17′) is found to be smaller than the electron diffraction value (118°24′). The deviations have been attributed to the various effects of non-rigidity.

7.2.18. AA′A″A‴BB′B″B‴ with C_{2v}-Symmetry

The NMR spectrum of 1,4-cyclohexadiene belongs to this group and has recently been investigated at 220 MHz [108]. The structural information deduced is not conclusive since the carbon skeleton may or may not be planar.

7.2.19. The 12-Spin-System with T_d-Symmetry

Dipolar couplings in tetrahedral compounds such as tetramethyl silane or neopentane should average zero even if the molecules are dissolved in nematic solvents. However, anisotropic pressure of the solvent may result in a slight distortion of the solute molecules producing non-zero values of the direct couplings. The proton spectrum of tetramethylsilane is shown in Fig. 32 [109]. Neopentane gives a similar spectrum.

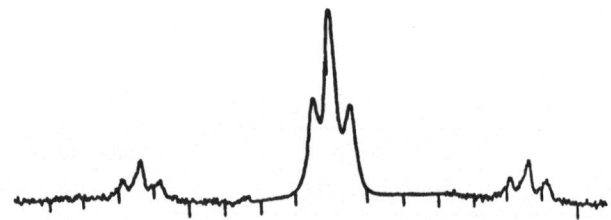

Fig. 32. Proton magnetic resonance spectrum of tetramethylsilane in the nematic phase of (III) [109]. Concentration = 26 mole%. Temperature = 75° C. The outer triplets are due to $^{13}C-H$ interactions at natural abundance of ^{13}C. The spectra were obtained from a 'CAT' (130 passes). The central triplet has been recorded at ($\frac{1}{32}$) of the gain for the satellites. The frequency markers are spaced at 10.6 Hz. (Reproduced from J. Chem. Phys. **44**, 4057 (1966), copyright (1966). Reprinted by permission of the copyright owner).

Intramethyl direct coupling is responsible for the observed triplet, whereas the intermethyl coupling contributes to line-width only. With the help of a computer of average transients (CAT), satellites due to $^{13}C-H$ interactions have also been observed (Fig. 32). Results obtained at various temperatures are summarised in Table 35.

Table 35. *Parameters obtained from the spectra of tetramethylsilane and neopentane [109] in the nematic phase of (III) (26 mole % solution)*

Compound	Temperature (°C)	D_{HH} (Hz)	$(J_{CH} + 2D_{CH})$ Hz
Tetramethylsilane	75	1.89	116.7
	80	1.40	118.64
	75	1.13	124.29
Neopentane	85	0.83	125.90
	95	–	127.19

D_{CH} and J_{CH} may be separated if (D_{HH}/D_{CH}) and J_{CH} are assumed to be temperature-independent.

The results correspond to changes in the $H-C-X$ angle by 0.1° (where X is Si or C) and/or in the CH_3 bond lengths of the order of 1 part in 10^3.

8. Determination of Absolute Signs of Indirect Coupling Constants

If the absolute sign of one of the parameters $(D_{ij}, J_{ij}, \Delta\sigma_i$ or S) affecting a spectrum in the nematic phase is known, that of the others can be derived. This can be illustrated from the spectra of fluoromethane [110]. Its proton and fluorine spectra in the isotropic phase consist of a doublet and a quartet respectively with $|J_{HF}| = 45 \pm 1$ Hz. The corresponding spectra in the nematic phase show a doublet of triplets and a quartet. Table 36 summarises the parameters obtained from the spectra.

Table 36. *Values of the parameters derived from the NMR spectra of fluoromethane in the nematic phase of (III) at 80° C [110]*

Inter-action	Splitting (Hz)	D_{HH} (Hz)	D_{HF} (Hz)	J_{HF} (Hz)	S_{zz}
H–H	525	+175			0.0175
H–F	234		–139.5	+45	0.0173
H–F	234		–94.5	–45	0.0117

In the nematic phase, the spectra are shifted downfield with respect to the isotropic medium. From a ^{19}F resonance study of CH_3F, trapped in a single crystal of β-quinol clathrate at $1.3°$ K, $\Delta\sigma_F$ is known to be negative. These results indicate that S_{zz} and hence D_{HH} must be positive and D_{HF} negative. Furthermore, since the values of S_{zz} obtained from D_{HH} or D_{HF} must be equal, the sign of J_{HF} is found to be positive (Table 36).

The absolute sign of the indirect $H - F$ couplings in fluoroform and methylene fluoride has been shown to be positive [140].

Signs of the indirect coupling constants between the protons of the two methyl groups in 2,4-hexadiyne [45] and the two methylene groups in allene [76] have similarly been shown to be positive and negative, respectively.

9. Spectra of Molecules Dissolved in a Nematic Phase Oriented by an Electric Field

Thus far the discussion has been restricted to nematic phases oriented by a magnetic field. However, the nematic phase, and consequently the molecules dissolved therein, may also be oriented by application of an electric field [29, 111 – 114, 132]. If the two fields are non-parallel the electric field causes the optic axis of the liquid crystal to rotate such that the angle (α) between the applied magnetic field and the optic axis becomes non-zero. As a consequence, the Hamiltonian (eq. 1) has to be modified. All the anisotropic contributions must be multiplied by $(^1/_2)$ $(3 \cos^2 \alpha - 1)$ [1, 31] giving relations (73) to (75):

$$D_{ij} = -\frac{h\gamma_i\gamma_j}{4\pi^2 r_{ij}^3} \cdot S_{ij} \cdot (^1/_2)(3\cos^2\alpha - 1) \tag{73}$$

$$
\begin{aligned}
P(\theta,\varphi) = \left(\frac{1}{4\pi}\right) &\{1 + (^1/_2)(3\cos^2\alpha - 1)(^5/_2)[3\cos^2\theta - 1]S_{zz} \\
&+ \sin^2\theta\cos 2\varphi\,(S_{xx} - S_{yy}) + 4\sin\theta\cos\theta\cos\varphi\cdot S_{xz} \\
&+ 4\sin\theta\cos\theta\sin\varphi\cdot S_{yz} + 2\sin^2\theta\,\mathrm{Sin}\,2\varphi\cdot S_{xy}]\}\,.
\end{aligned}
\tag{74}
$$

$$
\begin{aligned}
(\sigma_i + \sigma_{ia}) = (\tfrac{1}{3})(\sigma_{xxi} + \sigma_{yyi} + \sigma_{zzi}) &+ (^1/_2)(3\cos^2\alpha - 1)(\tfrac{2}{3})[(\sigma_{xxi}S_{xx} \\
&+ \sigma_{yyi}S_{yy} + \sigma_{zzi}S_{zz}) + S_{xz}(\sigma_{xzi} + \sigma_{zxi}) \\
&+ S_{yz}(\sigma_{yzi} + \sigma_{zyi}) + S_{xy}(\sigma_{xyi} + \sigma_{yxi})]\,.
\end{aligned}
\tag{75}
$$

The anisotropic contribution to the indirect coupling is obtained from (75) by substitution of J for σ and subtraction of $J_i = \frac{1}{3}(J_{xx} + J_{yy} + J_{zz})$. The remaining J_{ia} corresponds to $2D_{ij}^{ind}$.

Eqs. (73) to (75) show the importance of a variable angle α which changes the magnitudes and signs of the direct relative to the indirect couplings. Other anisotropic parameters are also affected. Such experiments not only help in the analysis of the spectra but also make it possible to obtain information generally lost as a result of the dominating influence of the direct couplings. A limitation, however, lies in the fact that all the parameters vary by the same factor (between $+1$ and -0.5) and hence no information on the absolute sign of orientation can be obtained.

Experiments have successfully been performed with an electric field applied along the sample axis perpendicular to the magnetic field [29]. Spectra of (cis) 1,2-dichloroethylene, furan, thiophene and benzene have been investigated in the nematic phase of (I) at 27°C.

For electric fields between 0 and 3 kV/cm, the spectra remain unaffected. Between 3.3 and 4.3 kV/cm, a fundamental change occurs. The lines at first broaden slightly without any shift. Then they fade and a broad peak appears near the centre (Fig. 33). This has been attributed to random orientation of the solute ($\alpha \approx 55°$ or $S \approx 0$). Further increase of the electric field results in spectra with normal line-width but reduced spread.

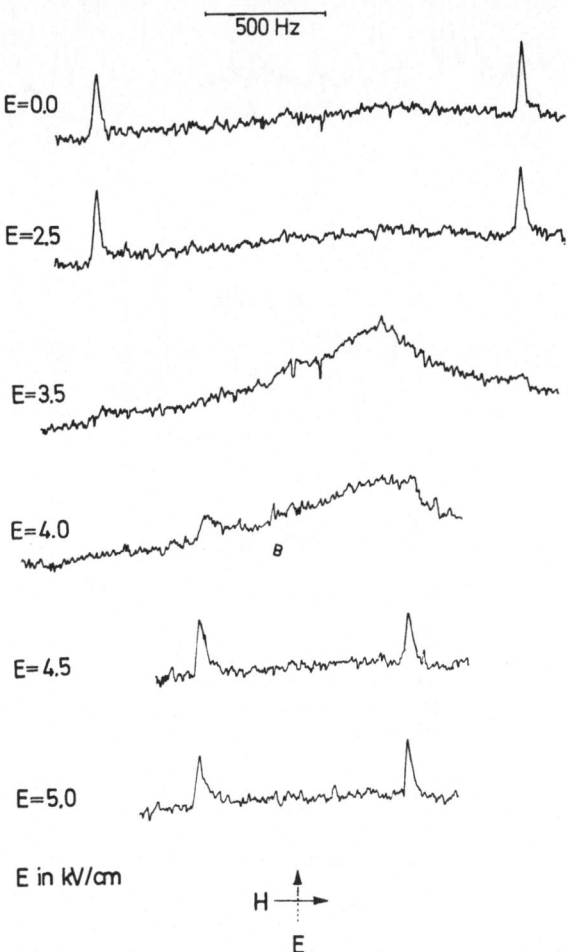

Fig. 33. Influence of electric fields on the NMR spectrum of (cis) 1,2-dichloroethylene dissolved in the nematic phase of (I) [29]. Concentration = 22 mole %. Temperature = 27°C. Spectrometer frequency = 60 MHz

6*

Results obtained (with and without the application of an electric field) from the spectral analysis of (cis) 1,2-dichloroethylene, furan, thiophene and benzene are given in Table 37. A typical spectrum of furan is shown in Fig. 34.

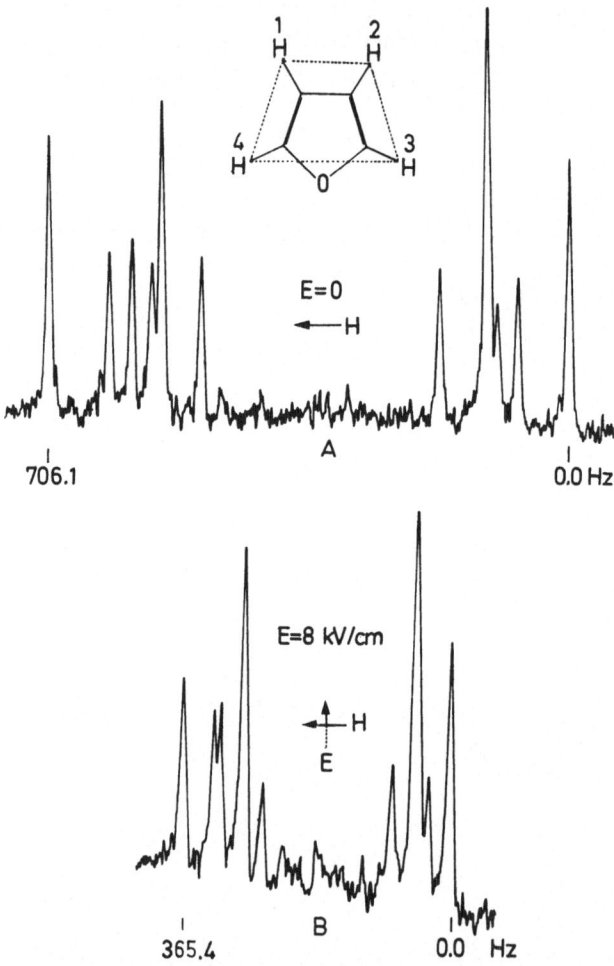

Fig. 34. NMR spectra of furan in the nematic phase of (I) [29]. (A) without electric field; (B) with electric field (8 kV/cm). Concentration of furan ≈ 24 mole%. Temperature = 27° C. Spectrometer frequency = 60 MHz

The observed variation in the direct couplings from D to $-(^1/_2)\,D$ (Table 37) has been attributed to the change in the angle α between the applied magnetic field and the optic axis of the liquid crystal from 0° to 90°. On application of the electric field, the molecules re-orient with their planes perpendicular to the direction of the magnetic field. This is further confirmed by the observed retention of orientation upon spinning of the sample.

Table 37. *Parameters derived from the spectra of (cis) 1,2-dichloroethylene furan, thiophene and benzene*

| Compound | Concentration (mole %) | Parameters obtained (Hz) | | | $\|D_{ij}^{E}/D_{ij}^{H}\|$ |
		Parameter	Without electric field	With electric field	
(cis) 1,2-di-chloroethylene	22	$\|D_{HH}\|$	575.0	288.0	0.50 ± 0.01
		D_{12}	-58.3	29.5	0.51 ± 0.01
		D_{23}	-163.7	81.3	0.51 ± 0.01
furan	≈ 24	D_{13}	-25.6	13.0	0.50 ± 0.01
		D_{34}	-19.6	8.9	0.45 ± 0.06
		$\nu_1 - \nu_3$	16.9 ± 0.4	16.4 ± 2.7	
		D_{12}	-35.6	18.3	0.51 ± 0.01
		D_{23}	-184.9	91.5	0.51 ± 0.01
thiophene	20	D_{13}	-19.9	10.2	0.49 ± 0.01
		D_{34}	-7.3	2.7	0.37 ± 0.13
		$\nu_1 - \nu_3$	72.5 ± 2	67.3 ± 5	
		D_{ortho}	-232.5	128.5	0.55 ± 0.01
benzene	24	D_{meta}	-44.6	23.8	0.53 ± 0.01
		D_{para}	-29.3	15.7	0.54 ± 0.01

The numbering in furan and thiophene corresponds to that given in Fig. 34. D_{ij}^{E} and D_{ij}^{H} are the values of the direct couplings observed with and without an electric field, respectively. Values of the indirect couplings used are identical to those of references [71, 53, 30].

10. Anisotropy of Chemical Shift

10.1. Introduction

NMR experiments in isotropic media provide only the average value or trace of the chemical shift tensor. In anisotropic solvents, however, it may be possible to obtain information on the anisotropy of the chemical shift. One experiment gives a single relation for the numerous tensor components. Therefore, measurements of shift anisotropies are more easily performed on molecules of high symmetry, for which the anisotropy is defined by comparatively few components.

For molecules with a 3-fold or higher axis of symmetry, the anisotropy can be obtained from the shift measurement as expressed in (76):

$$(\sigma_i + \sigma_{ia})_{\text{nematic}} - (\sigma_i)_{\text{isotropic}} = \tfrac{2}{3}(S_{zz} \cdot \Delta\sigma) \qquad (76)$$

where $\Delta\sigma = (\sigma_{\|} - \sigma_{\perp})$ and $\sigma_{\|}$ is measured along the symmetry axis.

For molecules with C_2-symmetry, the anisotropy is determined by eq. (77):

$$(\sigma_i + \sigma_{ia})_{\text{nematic}} - (\sigma_i)_{\text{isotropic}} = \tfrac{2}{3}(\sigma_{xxi} S_{xx} + \sigma_{yyi} S_{yy} + \sigma_{zzi} S_{zz}) \qquad (77)$$

10.2. Difficulties in the Measurement of Shift Anisotropies

Formulae (76) and (77) define the possibilities and limitations of the measurement of shift anisotropies. The sign and magnitude of $\Delta\sigma$ depend upon those of the orientation parameters and on differences of shifts. S-values in turn are derived from measured direct couplings and known inter-nuclear distances. The absolute signs of S-values are usually unknown but may be derived from the knowledge that solute molecules interact with the liquid crystal solvent mainly through dispersion forces. For $S > 0.5$, the absolute sign is positive by definition.

Measurements of shift differences introduce further complications in the determination of anisotropies:

(i) Measurements relative to an external reference require corrections for the anisotropy of bulk susceptibility as indicated by eq. (78) [115].

$$[\sigma_{\text{nem.}} - \sigma_{\text{iso.}}]_{\text{ext. ref.}} = (4\pi/9)\,\Delta\chi \qquad (78)$$

where $\Delta\chi = \chi_\| - \chi_\perp$. $\chi_\|$ and $\chi_\perp$ are the volume susceptibilities parallel and perpendicular to the field, respectively. A variation in the susceptibility due to volume change has been neglected. Since the anisotropy of the magnetic susceptibility of the liquid crystal is approximately 0.15×10^{-6} the necessary correction is about 0.2 ppm [116].

(ii) If the shifts are measured relative to internal standards, corrections are necessary to compensate for the changing influence of the anisotropy of the liquid crystal solvent on the internal standard upon transition from the isotropic to the nematic state. The magnitude of this effect may be estimated from benzene as a solvent which would shift an internal standard from 0.3 ppm to -0.3 ppm when going from the isotropic case to an orientation with the ring plane parallel to the magnetic field. This relative correction ($\Delta\sigma_{\text{loc.}} = -0.6$ ppm) may be smaller for the liquid crystal solvent but has been found to be in the vicinity of -0.4 to -0.5 ppm [117]. The necessary correction is:

$$[\sigma_{\text{nem.}} - \sigma_{\text{iso.}}]_{\text{int. ref.}} = \Delta\sigma_{\text{loc.}} \qquad (79)$$

Uncertainties in the shift measurement pointed out in (i) and (ii) have to be multiplied by the reciprocal of S in order to derive the uncertainty of $\Delta\sigma$. Thus in a case of threefold symmetry a precision in the shift measurement of ± 0.1 ppm and an orientation parameter (S_{zz}) of 0.01 result in a ± 15 ppm error in $\Delta\sigma$. This explains the large scatter observed in the anisotropy of the proton shift for fluoromethane, as obtained by several authors [110, 115, 117, 118] under different conditions of referencing.

It is obvious that measurements of the anisotropy in proton chemical shifts
are extremely problematic. The observed effects are of the same order of magnitude
as the corrections. For fluorine chemical shifts, on the other hand, the effects
are usually one order of magnitude larger than the corrections [116], but special
care is essential since the shifts are temperature dependent. This difficulty can
in principle be overcome by comparing the solute chemical shifts in nematic
solvents with and without spinning so that the isotropic-nematic shift is measured
without changing the temperature [140].

10.3. Results and Discussion

Shift anisotropy studies have thus far been restricted primarily to molecules
with 2-fold or greater axes of symmetry. They are summarised in Table 38 for
cases with a 3-fold or larger symmetry axis and Table 39 for molecules with
C_{2v}-symmetry.

Table 38. *Chemical shift anisotropies for molecules with a 3-fold or greater axis of symmetry*

Molecule	Nucleus studied	$(\sigma_{\parallel} - \sigma_{\perp}) \times 10^{-6}$	Remarks	Reference
benzene	H	-2.9 ± 0.2	the value is independent of temperature in a range of 30°C	[30, 3, 4]
1,3,5-trichloro-benzene	H	-4.91 ± 0.25		[3, 4]
1,3,5-tribromo-benzene	H	-7.41 ± 0.65		[119]
sym. triazine	H	3.89 ± 1.86		[119]
1,3,5-trifluoro-benzene	H	-2.6 ± 0.2		[7]
1,3,5-trifluoro-benzene	^{19}F	101 ± 5		[7]
hexafluoro-benzene	^{19}F	159		[31, 32]
hydrogen	H	-38 ± 6	uncorrected value	[120]
fluoromethane	H	-4.2 ± 1.5		[118]
fluoromethane	^{19}F	-157		[117, 110, 115]
fluoroform	H	9.8 ± 0.5		[140]
fluoroform	^{19}F	-83.2	with external reference	[140]
		-80.2 ± 2.0	with internal reference	
chloromethane	H	1.1 ± 0.9		[118]
bromomethane	H	1.3 ± 0.6		[118]
iodomethane	H	3.4 ± 0.4		[118]
iodomethane	^{13}C	-27.1 ± 2.9		[121]

Table 38 (continued)

Molecule	Nucleus studied	$(\sigma_{\parallel} - \sigma_{\perp}) \times 10^{-6}$	Remarks	Reference
trifluoro-acetic acid	^{19}F	235 ± 3	absolute sign determined (S > 0.5)	[122]
acetylene	H	8.1 ± 2.2	value for $\equiv$ CH	[57]
3-chloro-propyne	H	12.4	value for $\equiv$ CH	[45]
3-bromo-propyne	H	12.4	value for $\equiv$ CH	[45]
propyne	H	11.0 ± 1	value for $\equiv$ CH	[57]
1,3-pentadiyne	H	13.1 ± 0.3 for HC $\equiv$ -0.3 ± 0.5 for CH_3		[57]
3,3,3-trifluoro-propyne	^{19}F	-45.8 ± 2.3		[66]
perfluorodi-methyl acetylene	^{19}F	-51		[86]
ethane	H	3.4 ± 1.3		[141]
1,1,1-trifluoro-ethane	H	-4.1 ± 1.0		[141]
1,1,1-trifluoro-ethane	^{19}F	-70.0 ± 5.1		[141]
cyclobutadiene iron tricarbonyl	H	± 13		[80]
trichloromethyl silane	H	-26.8 ± 1.1		[123]
methyl cyanide	H	-0.76 ± 0.5		[141]
methyl isocyanide	H	-3.1 ± 0.5		[141]

Relatively few comparisons of experimental results with theoretical calculations have been made. The agreement is often poor. For the hydrogen molecule, for example, an uncorrected value of $\Delta\sigma$ has been found to be -24 to -44 ppm, whereas theory predicts a value between $+2$ to $+7$ ppm. For benzene and acetylene, on the other hand, the theoretical predictions for the anisotropy ($\Delta\sigma$) of -2.2×10^{-6} [135] and $\approx 10 \times 10^{-6}$ [40] respectively do not deviate considerably from the experimental results of $(-2.9 \pm 0.2) \times 10^{-6}$ [3, 4, 30] and $(8.1 \pm 2.2) \times 10^{-6}$ [57].

It has also been demonstrated that in methyl halides, the magnetic dipole model is inadequate for a description of the long range contribution to the shift anisotropies [123]. The model predicts anisotropy values of 2.5 to 9.8 ppm for the dipole at the nucleus and 2.7 to 20.8 ppm for the dipole at the midpoint of the bond, whereas the measured values vary from 1.1 to 3.4 ppm in going from chloride to iodide [123, 118].

The ^{19}F shielding anisotropy in 1,1,1-trifluoroethane indicates that the double bond character in the $C-F$ bond is of importance in determining the ^{19}F chemical shift anisotropy [141].

Table 39. *Shift anisotropy data for molecules with C_{2v}-symmetry*

Molecule	Nucleus	Axes	Result	Reference
fluoro-benzene	^{19}F	y-axis is the $C-F$ bond axis, z-axis is perpendicular to the plane of the ring	$(\sigma_{zz} - 0.076\,\sigma_{xx} - 0.924\,\sigma_{yy})$ $= 0.50 \times 10^{-4}$; when combined with C_6F_6 data, the anisotropy about the $C-F$ bond is $\sigma_{yy} - (^1/_2)(\sigma_{xx} + \sigma_{zz})$ $= 1.15 \times 10^{-4}$	[31]
pyrazine	H	x-axis along C_2-axis, z-axis perpendicular to the plane	$(\Delta\sigma = 2.45 \times 10^{-6}$ $-0.15\,\sigma_{xx})$	[74]
p-benzo-quinone	H	x-axis along C_2-axis, z-axis perpendicular to the plane	$(\Delta\sigma = -1.98 \times 10^{-6}$ $+1.63\,\sigma_{xx})$	[74]
methylene fluoride	H ^{19}F	z-axis is the intersection of the two perpendicular planes of symmetry, the x-axis lies in the plane containing protons and the y-axis in that with fluorines	$(\sigma_{zz})_H + 2.21\,(\sigma_{xx})_H$ $-(3.21\,\sigma_{yy})_H$ $= (5.93 \pm 1.6) \cdot 10^{-6}$ $(\sigma_{zz})_F + 2.21\,(\sigma_{xx})_F$ $-3.21\,(\sigma_{yy})_F$ $= (-280.9 \pm 1.0) \cdot 10^{-6}$	[140]

For more complex molecules, relations between the tensor components have been obtained (Table 39) [31, 74]. The individual components may in principle be isolated from different sites (in the same or different molecules) which are assumed to have the same degree of anisotropy. Isolation of the components is also possible from measurements in various nematic phases if the orientations are sufficiently different.

It has been suggested that the anisotropy be measured without reference to the isotropic state [29]. In this method the liquid crystal is subjected to an electric field along the sample axis. The field rotates the optic axis of the nematic phase 90° out of the magnetic field direction to an orientation parallel to the electric field. This reorientation results in a definite and predictable change in the S-values from $+A$ to $-(A/2)$ so that the variation observed in the chemical shift is 1.5 times the nematic-isotropic shift [29].

11. Anisotropy of the Indirect Coupling Constant

In molecules for which the direct couplings deviate from those predicted on the basis of known geometry and for which internal motions do not significantly influence the dipolar couplings, it is possible in principle to measure the anisotropy of the indirect couplings. As pointed out in section 4.5, relations for the anisotropic indirect couplings can be derived from eqs. (12) and (13) of section 4.4.

In this case again one measurement provides only one relation for the tensor components. Consequently measurements at different sites have to be combined.

Results obtained thus far are primarily qualitative. They have, therefore, been included in the relevant sections of chapter 7.

Recently, large contributions apparently due to anisotropic $^{13}C-H$ and $^{13}C-F$ indirect couplings in methyl fluoride enriched with 55% ^{13}C have been reported [139]. Either $(J_{\parallel} - J_{\perp})_{^{13}C-H}$ and $(J_{\parallel} - J_{\perp})_{^{13}C-F}$ are 1890 ± 130 Hz and 700 ± 130 Hz, respectively, or the geometry derived from the NMR data deviates significantly from that obtained from microwave and infrared studies.

12. Information from D and ^{35}Cl Magnetic Resonance Spectroscopy

12.1. Determination of Orientation and Structural Parameters

Deuteron magnetic resonance (DMR) spectroscopy in the nematic phase has been applied to orientation studies of the partially deuterated liquid crystalline $4-4'$-di(alcoxy)azoxy benzenes and alcoxy benzoic acids. Variations in the orientation parameters as a function of temperature and the side chain orientation of the liquid crystals have been measured [113].

From quadrupole coupling constants measured in crystalline powders at low temperatures and from DMR spectra in the nematic phase, degrees of orientation of the solute may be obtained [113]. Partially deuterated benzene, toluene, naphthalene and stilbene have thus been studied in nematic solvents. The $C-C-D$ bond angle in toluene-d_8 has been derived from the DMR spectrum. Splittings (Δv) of the CD_3 and the para $C-D$ deuterons are related by:

$$\{(\Delta v)_{CD_3}/(\Delta v)_{\text{para } C-D}\} = ((^3/_2) \cos^2 \lambda - (^1/_2)) \tag{80}$$

where λ is the angle between the two bond-directions.

Under the assumption that the field gradient tensor is rotationally symmetric relative to the $C-D$ bonds the angle λ in toluene-d_8 has been found to be $111.9°$.

12.2. Determination of Quadrupole Coupling Constants

Quadrupole coupling constants may be derived from the DMR spectra of partially deuterated molecules dissolved in the nematic phase if the values of the orientation parameters are known [38]. The S-values in turn are obtained from the proton spectra of the same sample at the same temperature, on the basis of known inter-proton distances. The precision of the measurement of a quadrupole coupling constant is consequently limited by the accuracy to which these distances are known.

The method has so far been applied to monodeuterobenzene [38] and partially deuterated acetonitrile [125]. The results are presented in Table 40.

Table 40. *Quadrupole coupling constants of deuterons (along the C−D bond) obtained from PMR and DMR spectra of partially oriented molecules*

Molecule	Distance used for the determination of S-Values	Quadrupole coupling constant	Reference
benzene-d_1	$(r_{HH})_{ortho} = 2.481 \pm 0.005$ Å	196.5 ± 1.3 KHz	[38]
acetonitrile-d_1 acetonitrile-d_2 acetonitrile-d_3	$(r_{HH}) = 1.803 \pm 0.001$ Å $(r_{C-H}) = 1.104 \pm 0.001$ Å	172.5 ± 1.5 KHz	[125]

A typical deuteron spectrum of monodeuterobenzene is presented in Fig. 35. The spectrum is only a doublet since the fine structure due to H−D interactions is not resolved.

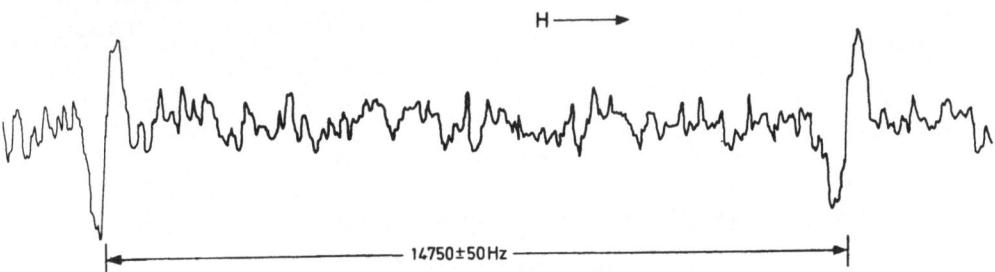

Fig. 35. DMR spectrum of monodeuterobenzene in the nematic phase of (I) + (II) [38]. Concentration = 24 mole%. Temperature = 27° C. Spectrometer frequency = 7.65 MHz

An isotopic effect on the degree of orientation has been observed. Orientation parameters are reduced by approximately 1% for each substituted deuteron. This limits the accuracy to which the quadrupole coupling constants can be measured in mixtures of deuterated with non-deuterated molecules of the compound, unless isotopic effects are understood quantitatively and corrections applied to the orientation parameters derived from non-deuterated species.

Solutions of poly-γ-benzyl-L-glutamate [132, 133] in a mixture of CH_2Cl_2 with CD_2Cl_2 as well as solutions of $CH_3CN + CD_3CN$, $CH_3Br + CD_3Br$, $CH_3I + CD_3I$ and $C_6H_6 + C_6D_6$ in the liquid crystalline nematic phases [138] have recently been used to determine the deuteron quadrupole coupling constants. Isotopic effects on orientation have been neglected.

In the first case, the two orientation parameters have been derived from known molecular geometry, from an assumed ^{35}Cl-quadrupole coupling constant of 72.4 MHz [134] and from observed splitting in the ^1H- and ^{35}Cl-spectra. A quadrupole coupling constant of 160 KHz has been obtained from the deuteron spectrum.

The constants summarised in Table 41 have been derived for CD_3I, CD_3Br, CD_3CN and C_6D_6.

Table 41. *D-quadrupole coupling constants measured along the C−D bond. Isotopic effects on orientation have been neglected* [138]

Molecule	Quadrupole coupling constant (KHz)
CD_3I	180 ± 5
CD_3Br	171 ± 4
CD_3CN	165 ± 5
C_6D_6	194 ± 4

The variations in quadrupole coupling constant in the substituted methanes (Table 41) have been attributed to the electron-withdrawing power of the substituents.

13. Papers which Appeared after the Manuscript was Submitted

13.1. The System AA′BB′XX′ with C_{2v}-Symmetry

^1H and ^{19}F spectra of ortho difluorobenzene have recently been studied [144]. The magnitudes and absolute signs of the indirect couplings have been found to be in agreement with those obtained from a study in isotropic media.

The geometrical information derived indicates displacement of the protons ortho to fluorine towards the fluorine atoms as compared to the "ideal" geometry (C-skeleton regular hexagon with $C-C = 1.390\,\text{A}°$, $C-H = 1.080\,\text{A}°$ and $C-F = 1.310\,\text{A}°$ and all bond angles equal to $120°$).

The anisotropy of the fluorine chemical shift in this molecule is described by eq. (81):

$$\sigma_{xx} - 0.653\,\sigma_{yy} - 0.347\,\sigma_{zz} = 123.9 \cdot 10^{-6} \qquad (81)$$

where x, y and z axes are perpendicular to the plane, along the $C-F$ bond and $\perp$ to the $C-F$ bond but in the molecular plane respectively.

13.2. The System ABB'CXX' with C_{2v}-Symmetry

A study of meta difluorobenzene [144] indicates that:
a) Magnitudes and signs of indirect couplings agree with those derived from "normal" NMR.
b) The derived geometry deviates from the "ideal" one (defined in section 13.1) in that the protons ortho to fluorine are displaced towards the fluorine atoms.
c) The anisotropy of the fluorine chemical shift is given by eq. (82)

$$\sigma_{xx} - 0.670\,\sigma_{yy} - 0.330\,\sigma_{zz} = 74.4 \cdot 10^{-6} \qquad (82)$$

where the axes are as defined in section 13.1.

Acknowledgements

The authors wish to thank the editors of the following journals for their permission to reproduce certain figures in the text: Angewandte Chemie, Canadian Journal of Chemistry, Journal of the American Chemical Society, Journal of Chemical Physics, Molecular Crystals, Molecular Physics, Österreichische Chemiker-Zeitung, Recueil des Travaux Chimiques des Pays-Bas, Science and Zeitschrift für Naturforschung. They are also grateful to Drs. R. A. BERNHEIM, P. J. BLACK, A. D. BUCKINGHAM, B. P. DAILEY, G. ENGLERT, G. R. LUCKHURST, C. MACLEAN, K. A. McLAUCHLAN, A. SAUPE, L. C. SNYDER and H. SPIESECKE for providing material prior to publication.

Appendix A

Liquid Crystal Solvents Used in NMR Spectroscopy

No.	Name	Structure	Nematic range (°C)
(I)	4-Methoxy benzylidene 4-amino-α-methyl cinnamic acid-n-propyl ester	$CH_3-O-\langle\text{ring}\rangle-CH=N-\langle\text{ring}\rangle-CH=C(CH_3)-C(=O)-OC_3H_7$	54 – 89
(II)	Anisole-p-azophenyl-n-capronate	$CH_3-O-\langle\text{ring}\rangle-N=N-\langle\text{ring}\rangle-O-C(=O)-C_5H_{11}$	66 – 106
(III)	4-4'-di-n-hexyloxy azoxy benzene	$C_6H_{13}-O-\langle\text{ring}\rangle-N=N(\rightarrow O)-\langle\text{ring}\rangle-O-C_6H_{13}$	81 – 127
(IV)	4-4'-di-n-heptyloxy azoxy benzene	$C_7H_{15}-O-\langle\text{ring}\rangle-N=N(\rightarrow O)-\langle\text{ring}\rangle-O-C_7H_{15}$	74 – 122.5
(V)	6-n-hexyloxy-2-napthoic acid	$H_{13}C_6O-\langle\text{naphthalene}\rangle-COOH$	147 – 198.5
(VI)	4-4'-di-n-hexyloxy azobenzene	$C_6H_{13}-O-\langle\text{ring}\rangle-N=N-\langle\text{ring}\rangle-O-C_6H_{13}$	102 – 114
(VII)	4-4'-di-n-hexyloxy benzalazine	$C_6H_{13}-O-\langle\text{ring}\rangle-C(H)=N-N=C(H)-\langle\text{ring}\rangle-OC_6H_{13}$	127 – 150
(VIII)	p-n-octyloxy benzoic acid dimer	$C_8H_{17}-O-\langle\text{ring}\rangle-C(=O\cdots H-O)(O-H\cdots O=)C-\langle\text{ring}\rangle-OC_8H_{17}$	108 – 147
(IX)	4-4'-di-methoxy azoxy benzene	$CH_3-O-\langle\text{ring}\rangle-N=N(\rightarrow O)-\langle\text{ring}\rangle-OCH_3$	118 – 136
(X)	4-4'-di-ethoxy azoxy benzene	$C_2H_5-O-\langle\text{ring}\rangle-N=N(\rightarrow O)-\langle\text{ring}\rangle-OC_2H_5$	138 – 168
(XI)	p-valeryloxy-p'-ethoxy azobenzene	$C_2H_5-O-\langle\text{ring}\rangle-N=N-\langle\text{ring}\rangle-O-C(=O)-C_4H_9$	79 – 125

Liquid Crystal Solvents Used in NMR Spectroscopy (continued)

No.	Name	Structure	Nematic range (°C)
(XII)	*p*-capronyloxy-*p'*-ethoxy azobenzene	$C_2H_5-O-\underset{}{\bigcirc}-N=N-\underset{}{\bigcirc}-O-\overset{O}{\underset{}{C}}-C_5H_{11}$	70 – 126
(XIII)	*o*-carbobutoxy-*p*-oxy-benzoic acid ethoxy phenyl ester	$C_2H_5-O-\underset{}{\bigcirc}-O-\underset{O}{\overset{}{C}}-\underset{}{\bigcirc}-O-\overset{O}{\underset{}{C}}-OC_4H_9$	54 – 67
(XIV)	4-4'-di-*n*-butoxy azoxy-benzene	$C_4H_9-O-\underset{}{\bigcirc}-\underset{O}{N=N}-\underset{}{\bigcirc}-OC_4H_9$	107 – 134
(XV)	4-4'-di-*n*-pentoxy azoxybenzene	$C_5H_{11}-O-\underset{}{\bigcirc}-\underset{O}{N=N}-\underset{}{\bigcirc}-OC_5H_{11}$	82 – 119

Appendix B

Compounds Studied

Spectral type	Reference	Nucleus studied	Compounds
AB	[51] [60]	^{1}H ^{1}H	4,6-dichloropyrimidine; 2,4,5-trichloronitrobenzene
AX	[4]	^{1}H	1,2,4,5-tetrachlorofluoro-benzene
A$_2$	[45, 28, 4, 23, 29, 120, 132, 133]	^{1}H D ^{35}Cl	acetylene; 1,2,3,5-tetrachloro-benzene; 1,2,4,5-tetrachloro-benzene; (cis)-(trans)- and (gem)-dichloroethylenes; methylene chloride-d$_2$; hydrogen and methylene chloride
ABC	[61]	^{1}H	3,3,3-trichloropropylene oxide (racemic)
≈AA'A''	[50]	^{1}H	1,2,4-trichlorobenzene
AB$_2$	[35, 45, 57, 44]	^{1}H	3-chloropropyne; 3,5-di-chlorobenzoic acid; 3-bromo-propyne; 1,2,3-trichloro-benzene
≈AA'$_2$	[50]	^{1}H	2,6-dibromopyridine
A$_3$	[4, 57, 63, 28, 25,119, 122, 118, 125]	^{1}H D	1,3,5-trichlorobenzene; methylcyanide; methyl-alcohol; methyliodide; sym. triazine; methylchloride; tri-fluoroacetic acid; CH$_3$CN; CDH$_2$CN; CD$_2$HCN and CD$_3$CN
AB$_3$ (without symmetry)	[44]	^{1}H	2,3,4,6- and 2,3,5,6-tetra-chloroanisole
A$_3$B$_2$	[41, 26]	^{1}H	ethyliodide and ethylalcohol

Compounds Studied (continued)

Spectral type	Reference	Nucleus studied	Compounds
AB_3 and AX_3 (with C_3-symmetry)	[44, 57, 64, 65, 66, 110, 115, 117, 118, 140, 141, 143]	^{1}H	2,3,5,6-tetrachlorotoluene; propyne; 1,3-pentadiyne; $^{13}CH_3OH$; $^{13}CH_3I$; $^{13}CH_3CN$; $CH_3$$^{13}CN$; $CH_3C$$^{15}N$; $^{13}CH_3NC$; 3,3,3-trifluoropropyne; fluoromethane; fluoroform; methyl mercuric chloride
AA'BB' (C_{2v}-symmetry)	[50, 71–73, 29, 124]	^{1}H	furan; thiophene; benzo-furazan oxide; symmetrical orthodisubstituted benzenes; pyridazine
AA'A"A''' (D_{2d}-symmetry)	[75, 76]	^{1}H	allene
A_2X_2	[140]	^{1}H, ^{19}F	methylenefluoride
AA'XX' (C_{2v}-symmetry)	[77, 78]	^{1}H, ^{19}F	1,1-difluoroethylene
AA'A"A''' (D_{2h}-symmetry)	[35, 48, 58, 74, 76, 81, 77]	^{1}H ^{19}F	p-dichlorobenzene; p-di-bromobenzene; p-diiodo-benzene; ethylene; tetra-fluoroethylene; p-benzo-quinone; pyrazine; cyclo-butadiene iron tricarbonyl; ethylene oxide and ethylene sulphide
AB_2C (C_{2v}-symmetry)	[59]	^{1}H	m-dichloro and m-dibromo benzene
AA'BB'C (C_{2v}-symmetry)	[55]	^{1}H	pyridine
AA'BB'X (C_{2v}-symmetry)	[60]	^{1}H	p-chloro and p-bromo fluorobenzene
AA'A"A'''A''''A''''' with C_6-symmetry	[15, 27, 30, 7, 32, 33, 29]	^{1}H ^{19}F	benzene and hexafluoro-benzene

Compounds Studied (continued)

Spectral type	Reference	Nucleus studied	Compounds
$A_3A_3'(A_3X_3)$ with C_3-symmetry	[45, 86, 76, 136, 141]	^1H ^{19}F	2,4-hexadiyne; dimethyl acetylene; perfluorodimethyl acetylene; ethane; 1,1,1-tri-fluoroethane and mercury dimethyl
AA'A''A'''A''''A''''' (D_{3h}-symmetry)	[10, 34, 92, 127]	^1H	cyclopropane and s-trioxane
AA'A''XX'X'' (D_{3h}-symmetry)	[94, 95, 7]	^1H ^{19}F	1,3,5-trifluorobenzene
AA'A''A'''XX' with D_{2h}-symmetry	[48]	^1H	p-difluorobenzene and symmetrical tetrafluoro-benzene
AA'BB'CC' with C_{2v}-symmetry	[10, 99]	^1H	bicyclobutane
AA'BB'CX with C_{2v}-symmetry	[31, 38]	^1H ^{19}F, D	fluorobenzene and mono-deuterobenzene
AA'BB'XX' with C_{2v}-Symmetry	[144]	^1H	o-difluorobenzene
ABB'CXX' with C_{2v}-symmetry	[144]	^1H	m-difluorobenzene
A_3B_2X with C_s-symmetry	[103]	^1H ^{19}F	ethylfluoride
AA'A''A'''A''''A'''''A''''''A''''''' with D_{2d}-symmetry	[34, 10, 126]	^1H	cyclobutane and spiropentane
AA'A''A'''BB'B''B''' with C_{2v}-symmetry	[108]	^1H	1,4-cyclohexadiene
$(A_3)_4$ with T_d-symmetry	[109]	^1H	tetramethylsilane and neopentane

References

1. BUCKINGHAM, A. D., and J. A. POPLE: Trans. Faraday Soc. 59, 2421 (1963) and references therein.
2. DUCROS, P.: Bull. Soc. Franc. Mineral. Crist. 83, 85 (1960).
3. SAUPE, A., and G. ENGLERT: Phys. Rev. Letters 11, 462 (1963).
4. ENGLERT, G., and A. SAUPE: Z. Naturforsch. 19a, 172 (1964).
5. BUCKINGHAM, A. D., and K. A. McLAUCHLAN: Progress in NMR spectroscopy, Vol. 2, edited by J. W.EMSLEY, J. FEENEY, and L. H. SUTCLIFFE, Oxford: Pergamon Press 1967.
6. SAUPE, A.: Angew. Chemie (Int. Ed.) 7, 107 (1968).
7. SAUPE, A.: Angew. Chemie 80, 99 (1968).
8. LUCKHURST, G. R.: Quarterly Reviews 22, 179 (1968).
9. LUCKHURST, G. R.: Österr. Chemiker-Ztg. 4, 113 (1967).
10. MEIBOOM, S., and L. C. SNYDER: Science 162, 1337 (1968).
11. GRAY, G. W.: Molecular structure and properties of liquid crystals. New York: Academic Press Inc. 1962.
12. FRIEDEL, M. G.: Ann. Physique 18, 273 (1922).
13. FRIEDEL, M. G.: Compt. Rend. 176, 475 (1923).
14. CANO, M. R.: Compt. Rend. 251, 1139 (1960).
15. SACKMANN, E., S. MEIBOOM, and L. C. SNYDER: J. Am. Chem. Soc. 89, 5981 (1967).
16. BOSE, E.: Z. Physik 10, 32 (1909).
17. ZOCHER, H.: Trans. Faraday Soc. 29, 945 (1933).
18. LUCKHURST, G. R.: Mol. Cryst. 2, 363 (1967).
19. LANDOLT-BÖRNSTEIN: Zahlenwerte und Funktionen aus Physik, Chemie, Astronomie, Geophysik und Technik, Bd. II, Teil 2a, S. 266. Berlin-Göttingen-Heidelberg: Springer 1960.
20. SPIESECKE, H., and J. B. JOURDAN: Angew. Chemie (Int. Ed.) 6, 450 (1967).
21. DEMUS, D.: Z. Naturforsch. 22a, 285 (1967).
22. PANAR, M., and W. D. PHILLIPS: J. Am. Chem. Soc. 90, 3880 (1968).
23. SAMULSKI, E., and A. V. TOBOLSKY: Mol. Cryst. 7, 433 (1969).
24. SOBAJIMA, S.: J. Phys. Soc. Japan 23, 1070 (1967).
25. LAWSON, K. D., and T. J. FLAUTT: J. Am. Chem. Soc. 89, 5489 (1967).
26. BLACK, P. J., K. D. LAWSON, and T. J. FLAUTT: Mol. Cryst. 7, 201 (1969).
27. BLACK, P. J., K. D. LAWSON, and T. J. FLAUTT: J. Chem. Phys. 50, 542 (1969).
28. DIEHL, P., and C. L. KHETRAPAL: Mol. Phys. 14, 283 (1967).
29. DIEHL, P., C. L. KHETRAPAL, H. P. KELLERHALS, U. LIENHARD, and W. NIEDERBERGER: J. Mag. Res. 1, 527 (1969).
30. SAUPE, A.: Z. Naturforsch. 20a, 572 (1965).
31. SNYDER, L. C.: J. Chem. Phys. 43, 4041 (1965).
32. SNYDER, L. C., and E. W. ANDERSON: J. Chem. Phys. 42, 3336 (1965).
33. SNYDER, L. C., and E. W. ANDERSON: J. Am. Chem. Soc. 86, 5023 (1964).
34. MEIBOOM, S., and L. C. SNYDER: J. Am. Chem. Soc. 89, 1038 (1967).
35. SAUPE, A.: Mol. Cryst. 1, 527 (1966).
36. NEHRING, J., and A. SAUPE: Mol. Cryst. 8, 403 (1969).
37. HERSCHBACH, D. R., and V. W. LAURIE: J. Chem. Phys. 37, 1668 (1962).
38. DIEHL, P., and C. L. KHETRAPAL: Can. J. Chem. 47, 1411 (1969).
39. DIEHL, P. and H. P. KELLERHALS: J. Mag. Res. 1, 196 (1969).
40. POPLE, J. A., W. G. SCHNEIDER, and H. J. BERNSTEIN: High Resolution Nuclear Magnetic Resonance. McGraw Hill Book Company, Inc. 1959.
41. WOODMAN, C. M.: Mol. Phys. 13, 365 (1967).
42. MUSHER, J. I.: J. Chem. Phys. 46, 1537(1967).
43. SAUPE, A., and J. NEHRING: J. Chem. Phys. 47, 5459 (1967).
44. ENGLERT, G., and A. SAUPE: Z. Naturforsch. 20a, 1401 (1965).
45. ENGLERT, G., A. SAUPE, and J. P. WEBER: Z. Naturforsch. 23a, 152 (1968).
46. DIEHL, P.: Helv. Chim. Acta 48, 567 (1965).

47. DIEHL, P., and D. TRAUTMANN: Mol. Phys. **11**, 531 (1966).
48. BULTHUIS, J., J. GERRISTEN, C. W. HILBERS, and C. MACLEAN: Rec. Trav. Chim. **87**, 417 (1968).
49. NEHRING, J.: Ph. D. Thesis, Freiburg University 1968.
50. DIEHL, P., C. L. KHETRAPAL, and U. LIENHARD: Mol. Phys. **14**, 465 (1968).
51. DIEHL, P., and H. P. KELLERHALS: (unpublished).
52. HEWITT, R. C.: Rev. Scient. Instr. **39**, 1066 (1968).
53. DIEHL, P., C. L. KHETRAPAL, and U. LIENHARD: Can. J. Chem. **46**, 2645 (1968).
54. SNYDER, L. C., and R. L. KORNEGAY: American Chemical Soc. Symposium on ordered fluids and liquid crystals. Atlantic City, New Jersey 1965.
55. DIEHL, P., C. L. KHETRAPAL, and H. P. KELLERHALS: Mol. Phys. **15**, 333 (1968).
56. CASTELLANO, S., and A. A. BOTHNER-BY: J. Chem. Phys. **41**, 3863 (1964).
57. ENGLERT, G., and A. SAUPE: Mol. Cryst. **1**, 503 (1966).
58. DIEHL, P., and C. L. KHETRAPAL: (unpublished).
59. DIEHL, P., and C. L. KHETRAPAL: Mol. Phys. **15**, 633 (1968).
60. DIEHL, P., C. L. KHETRAPAL, and U. LIENHARD: Org. Mag. Res. **1**, 93 (1969).
61. SACKMAN, E., S. MEIBOOM, and L. C. SNYDER: J. Am. Chem. Soc. **90**, 2183 (1968).
62. WHITMAN, D. R.: J. Mol. Spect. **10**, 250 (1962).
63. SAUPE, A., G. ENGLERT, and A. POVH: Advan. Chem. Ser. **63**, 51 (1967).
64. ENGLERT, G., and A. SAUPE: Mol. Cryst. **8**, 233 (1969).
65. SPIESECKE, H.: Z. Naturforsch. **23a**, 467 (1968).
66. BUCKINGHAM, A. D., E. E. BURNELL, C. A. DE LANGE, and A. J. REST: Mol. Phys. **14**, 105, (1968).
67. COSTAIN, C. C.: J. Chem. Phys. **29**, 864 (1958).
68. KESSLER, M., H. RING, R. TRAMBARULO, and W. GORDY: Phys. Revs. **79**, 54 (1950).
69. VENKATESHWARLU, P., and W. GORDY: J. Chem. Phys. **23**, 1200 (1955).
70. MILLER, S. L., L. C. AAMODT, G. DOUSMANIS, C. H. TOWNES, and J. KRAITCHMAN: J. Chem. Phys. **20**, 1112 (1952).
71. DIEHL, P., C. L. KHETRAPAL, and H. P. KELLERHALS: Helv. Chim. Acta **51**, 529 (1968).
72. DIEHL, P., and C. L. KHETRAPAL: Mol. Phys. **15**, 201 (1968).
73. DIEHL, P., and C. L. KHETRAPAL: Proc. XV Collôque Ampère, Grenoble 1968, p. 251.
74. DIEHL, P., and C. L. KHETRAPAL: Mol. Phys. **14**, 327 (1968).
75. SPIESECKE, H., and J. B. JOURDAN: Angew. Chem. **79**, 475 (1967).
76. SPIESECKE, H.: (private communication).
77. SAUPE, A., and H. SPIESECKE: Mol. Cryst. (in the press).
78. BUCKINGHAM, A. D., E. E. BURNELL, and C. A. DE LANGE: Mol. Phys. **16**, 299 (1969).
79. EDGELL, W. F., P. A. KINSEY, and J. W. AMY: J. Am. Chem. Soc. **79**, 2691 (1957).
80. YANNONI, C. S., G. P. CEASAR, and B. P. DAILEY: J. Am. Chem. Soc. **89**, 2833 (1967).
81. BOVÉE, W., C. W. HILBERS, and C. MACLEAN: Mol. Phys. **17**, 75 (1969).
82. CASTELLANO, S., C. SUN, and R. KOSTELNIK: J. Chem. Phys. **46**, 327 (1967).
83. BAK, B., L. HANSEN, and J. R. ANDERSEN: J. Mol. Spect. **2**, 361 (1958).
84. MARTIN, T. E., and A. H. KALANTAR: J. Chem. Phys. **49**, 235 (1968).
85. LONGUET-HIGGINS, H. C.: Mol. Phys. **6**, 445 (1963).
86. BUCKINGHAM, A. D., E. E. BURNELL, and C. A. DE LANGE: Mol. Phys. **15**, 285 (1968).
87. PIGNATARO, E., and B. POST: Acta Cryst. **8**, 672 (1955).
88. THOMAS, L. F., E. I. SHERRARD, and J. SHERIDAN: Trans. Faraday Soc. **51**, 619 (1955).
89. TRAMBARULO, R., and W. GORDY: J. Chem. Phys. **18**, 1613 (1950).
90. SHEEHAN, W. F., and V. SCHOMAKER: J. Am. Chem. Soc. **74**, 4468 (1952).
91. SHOOLERY, J. N., R. G. SHULMAN, W. F. SHEEHAN, V. SCHOMAKER, and D. M. YOST: J. Chem. Phys. **19**, 1364 (1951).
92. SNYDER, L. C., and S. MEIBOOM: J. Chem. Phys. **47**, 1480 (1967).
93. BASTIANSEN, O., F. N. FRITSCH, and K. HEDBERG: Acta Cryst. **17**, 538 (1964).
94. YIM, C. T., and D. F. R. GILSON: Can. J. Chem. **46**, 2783 (1968).
95. LASZLO, P.: Prog. NMR Spectroscopy **3**, 321 (1967).
96. JONES, R. G., R. C. HIRST, and H. J. BERNSTEIN: Can. J. Chem. **43**, 683 (1965).
97. OOSAKA, H., A. SEKINE, and T. SAITO: Bull. Chem. Soc. Japan **27**, 182 (1954).

98. Bak, B., D. Christensen, L. H. Nygaard, and E. Tannenbaum: J. Chem. Phys. **26**, 134 (1957).
99. Snyder, L. C., and S. Meiboom: Mol. Cryst. **7**, 181 (1969).
100. Langseth, A., and B. P. Stoischeff: Can. J. Phys. **34**, 350 (1956).
101. Nygaard, L., I. Bojesen, T. Pedersen, and J. R. Andersen: J. Mol. Structure **2**, 209 (1968).
102. Laurie, V. W., and D. R. Herschbach: J. Chem. Phys. **37**, 1687 (1962).
103. Buckingham, A. D., E. E. Burnell, and C. A. De Lange: Mol. Phys. **16**, 191 (1969).
104. Kraitchman, J., and B. P. Dailey: J. Chem. Phys. **23**, 184 (1955).
105. Bak, B., S. Detoni, L. H. Nygaard, J. T. Nielsen, and J. R. Andersen: Spectrochim. Acta **16**, 376 (1960).
106. Nygaard, L.: Spectrochim. Acta **22**, 1261 (1966).
107. Almenningen, A., O. Bastiansen, and P. N. Skancke: Acta Chem. Scand. **15**, 711 (1961).
108. Buckingham, A. D., E. E. Burnell, and C. A. De Lange: Mol. Phys. **16**, 521 (1969).
109. Snyder, L. C., and S. Meiboom: J. Chem. Phys. **44**, 4057 (1966).
110. Bernheim, R. A., and B. J. Lavery: J. Am. Chem. Soc. **89**, 1279 (1967).
111. Twitchel, R. P., and E. F. Carr: J. Chem. Phys. **46**, 2756 (1967).
112. Meier, G., and A. Saupe: Mol. Cryst. **1**, 515 (1966).
113. Rowell, J. C., W. D. Phillips, L. R. Melby, and M. J. Panar: J. Chem. Phys. **43**, 3442 (1965).
114. Chen, D. H., and G. R. Luckhurst: Mol. Phys. **16**, 91 (1969).
115. Buckingham, A. D., and E. E. Burnell: J. Am. Chem. Soc. **89**, 3341 (1967).
116. Buckingham, A. D., E. E. Burnell, and C. A. De Lange: J. Am. Chem. Soc. **90**, 2972 (1968).
117. Bernheim, R. A., and T. R. Krugh: J. Am. Chem. Soc. **89**, 6784 (1967).
118. Caesar, G. P., C. S. Yannoni, and B. P. Dailey: J. Chem. Phys. **50**, 373 (1969).
119. Yannoni, C. S., and B. P. Dailey: (private communication).
120. Buckingham, A. D., E. E. Burnell, and C. A. De Lange: Chem. Comm. 1408 (1968).
121. Yannoni, C. S., and E. B. Whipple: J. Chem. Phys. **47**, 2508 (1967).
122. Dunn, M. B.: Mol. Phys. **15**, 433 (1968).
123. Caesar, G. P., and B. P. Dailey: (private communication).
124. Burnell, E. E., and C. A. De Lange: Mol. Phys. **16**, 95 (1969).
125. Diehl, P., and C. L. Khetrapal: J. Mag. Res. **1**, 524 (1969).
126. Buckingham, A. D., E. E. Burnell, and C. A. De Lange: Mol. Phys. **17**, 205 (1969).
127. Cocivera, M.: J. Chem. Phys. **47**, 3061 (1967).
128. Maryott, A. A., and S. F. Acree: J. Res. Nat. Bur. Std. **33**, 71 (1944).
129. Kimura, M., and K. Aoki: J. Chem. Soc. Japan **72**, 169 (1951).
130. Shand, W.: Acta Cryst. **3**, 54 (1950).
131. Oka, T., K. Tuschiya, S. Iwata, and Y. Morion: Bull. Chem. Soc. Japan, **37**, 4 (1964).
132. Gill, D., M. P. Klein, and G. Kotowycz: J. Am. Chem. Soc. **90**, 6870 (1968).
133. Klein, M. P., D. Gill, and G. Kotowycz: Chem. Phys. Lett. **2**, 677 (968).
134. Livingston, R. L.: J. Chem. Phys. **19**, 1434 (1951).
135. Waugh, J. S., and R. W. Fessenden: J. Am. Chem. Soc. **79**, 846 (1957).
136. Englert, G.: Z. Naturforsch. **24a**, 1074 (1969).
137. Bernheim, R. A., and B. J. Lavery: J. Colloid and Interface Sci. **26**, 291 (1968).
138. Caspary, W. J., F. Millet, M. Reichbach, and B. P. Dailey: (private communication).
139. Krugh, T. R., and R. A. Bernheim: J. Am. Chem. Soc. **91**, 2385 (1969).
140. Bernheim, R. A., D. J. Hoy., T. R. Krugh, and B. J. Lavery: J. Chem. Phys. **50**, 1350, (1969).
141. Silverman, D. N., and B. P. Dailey: (private communication).
142. Maier, W., and A. Saupe: Z. Naturforsch. **13a**, 564 (1958); **14a**, 882 (1959); **15a**, 287 (1960).
143. Ayres, M., K. A. McLauchan, and J. Wilkinson: Chem. Comm. 858 (1969).
144. Yim, C. T., and D. F. R. Gilson: Can. J. Chem. **47**, 1057 (1969).

The Use of Symmetry in Nuclear Magnetic Resonance

R. G. JONES

Chemistry Department
University of Essex, Colchester, England

Contents

1. Introduction 100
  1.1. Symmetry and Group Theory 100
  1.2. The Role of Symmetry in Spectroscopy 101
  1.3. Symmetry in Nuclear Magnetic Resonance 101

2. Symmetry and Symmetry Groups of Rigid and Non-rigid Systems 102
  2.1. Symmetry Elements 102
  2.2. Symmetry Operations 103
  2.3. Symmetry Classification of Rigid Systems 104
  2.4. Symmetry of Non-rigid Molecules 106

3. The NMR Permutation Group and Symmetry 108
  3.1. Symmetry and the NMR Hamiltonian 108
  3.2. Isolated Groups of Nuclei (Same Isotopic Species) 109
    a.1. The Example of Methane in an Isotropic Medium 110
    a.2. The Example of Methane in an Anisotropic Medium 110
    a.3. Groups of Nuclei with Spin $\geqslant 1$ 110
    b.1. The Example of Ethylene in an Isotropic Medium 111
    b.2. The Example of Ethylene in an Anisotropic Medium 111
  3.3. Interacting Groups of Spin 1/2 Nuclei 111
    The System $A_mB_nC_p$... 112
    $AA'X_nX'_n$ and $AA'B_nB'_n$ Systems 112
    The $A_2A'_2BB'B''B'''$ and $A_2A'_2XX'X''X'''$ Systems 113
    Systems Containing Spin $\geqslant 1$ Nuclei 115
    The Effect of Time-averaging Processes on the Hamiltonian and Permutation Group 117

4. Factorisation of the Determinant of the Hamiltonian Matrix 118
  4.1. Total Magnetic Quantum Number Factorisation 118
  4.2. Total Spin Quantum Number Factorisation 119
    4.2.1. Isolated Groups of Nuclei (Spin I = 1/2) 120

4.2.2. Interacting Groups of Nuclei 120
4.2.3. Total Spin and Symmetry 121
4.3. Symmetry Factorisation of the Determinant of the Hamiltonian
Matrix 122
The AA'BB' System-o-Dichlorobenzene 123
The AA'A″BB'B″ System 127
AA'A″A‴BB'B″B‴ Systems 131

5. The X Approximation..................................... 133
5.1. Homonuclear Systems of Nuclei 134
5.2. Heteronuclear Systems of Nuclei 135
5.3. Sub-spectral Analysis.................................. 140
The AA'XX' System 140
The AA'A″XX'X″ System 140
The AA'A″A‴XX'X″X‴ Systems 141

Matrices and Vectors (Appendix A) 142
The Symmetry Group, Character Table Determinant Factorisation and
Symmetrised Wavefunctions for the $A_2A_2'XX'X''X'''$ System (Appendix B) 148
Group Character Tables (Appendix C)............................ 164
References 174

Glossary of Symbols

$[1]$, $[20]$	References, italicised.
[1], [57]	Abbreviated notation for spin-wavefunctions which locates the β's.
$2[1]-[2]-[3]$	Linear symmetrised combination of spin-wavefunctions, not normalised.
(123), (57) (68)	Permutations $(1 \to 2 \to 3 \to 1)$, $(5 \leftrightarrow 7)$ $(6 \leftrightarrow 8)$.
A, B, X, ...	Symmetry elements or operations.
E	Identity operation.
P	Permutation, (12)
E*	Inversion of all particles through a centre of mass.
P*	Permutation inversion (12)(45)*.
A, B, E,	Symmetry species or irreducible representations, italicised.
Γ	Representation of the group.
C_{nv}, D_{nh}, G	Symmetry Point Groups, italicised.
P_n	Permutation Point Group, italicised.
σ	Plane of symmetry.
$\mathbf{C}_n$	Rotation axis of symmetry, (bold sub-script).
i	Centre of symmetry.
$\mathbf{S}_n$	Rotation-reflection axis of symmetry, (bold sub-script).
h	Order of mathematical group.
l_i	Dimension of the ith representation = order of the matrices which constitute it.

C_i	Class of mathematical group.
χ_c	Number of spin-wavefunctions of a given set which remain unchanged under operations in class C.
g_c	Number of operations in class C.
$\left.\begin{array}{l}\chi_c^{(\gamma)} \\ \chi_p^{(\gamma)} \\ \chi(P)^{\Gamma_i}\end{array}\right\}$	Symmetry group characters, corresponding to the class C or permutation P and the symmetry species γ or the representation Γ_i.
v_A, v_a	Larmor frequency and effective Larmor frequency.
J_{AB}, J_{ab}	Scalar coupling constant and effective coupling constant.
D_{AB}	Direct coupling constant.
$I^2, I_X, I_Y, I_Z, I_i, I_j$	Spin operators.
H, H^J, H^D	Hamiltonians.
$A_3B_2, AA'X_nX'_n$	NMR spin system notation, primes denote magnetic non-equivalence.
$[AX_n]_2$	Square brackets denote magnetic non-equivalence.
ab, aa'bb'	NMR sub-spectral notation (primes as above).
φ, ψ	Spin wavefunctions.
m	Magnetic quantum number, eigenvalue of I_z.
m_T	Total magnetic quantum number, sum of m taken over a complete system.
$m(XX'), m_X$	Sum of m taken over a group of nuclei.
I_T	Total spin quantum number, associated with I^2.
E_n	Energy level.
$\hbar$	$h/2\pi$ Planck's constant over two pi-unit of angular momentum.
$\vec{r}, \vec{v}$	Vectors.
$\vec{i}, \vec{j}, \vec{k}$	Unit vectors.
x, y, z	Cartesian co-ordinates.
M_1, M_2	Matrices, italicised.
m_{ij}	Elements of a matrix in the ith row and j^{th} column.
ε	Unit matrix.

1. Introduction

It has been the constant quest of the chemist to simplify the calculation of molecular parameters and to understand the theory underlying such calculations. The recognition of symmetry in its broadest sense and its relationship with group theory bring to hand information essential to the understanding and execution of fundamental calculations at a personal level. This approach also leads to a fuller comprehension of the results derived from the computer facility now available to all.

It is the purpose of this paper to present the essential details necessary for the fullest simplification of the problems met in high resolution nuclear magnetic resonance spectroscopy using the symmetry properties of the molecule or system. It is appropriate, however, to begin with a rather more general discussion of the implications of symmetry.

1.1. Symmetry and Group Theory

Symmetry to many is a visual concept manifested as regular geometrical shapes possessed by molecules such as benzene (regular hexagon) or even snowflakes! Symmetry can however be shown to consist of a number of elements which may include a plane, an axis a centre of inversion or special combinations of these.

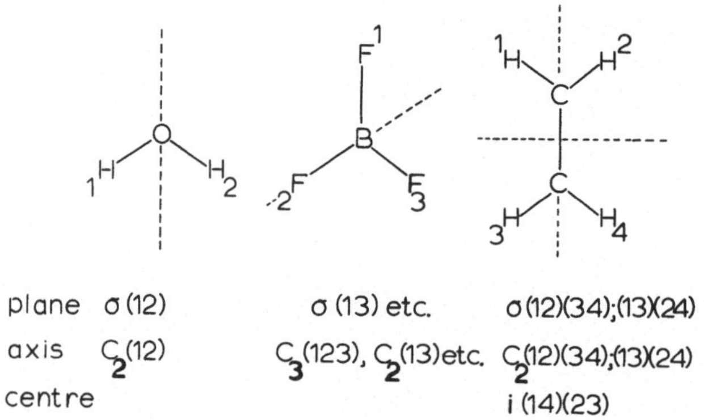

plane $\sigma(12)$ $\sigma(13)$ etc. $\sigma(12)(34);(13)(24)$

axis $C_2(12)$ $C_3(123)$, $C_2(13)$etc. $C_2(12)(34);(13)(24)$

centre $i(14)(23)$

Fig. 1.1. Basic elements of symmetry and equivalent permutations of nuclei

The symmetry elements impress the eye but it will become apparent that their importance in theoretical problems arises from the implied reflection, exchange or inversion of equivalent features of the system. It is possible in fact to replace the elements of symmetry in a given system by a notation which implies the exchange or permutation of equivalent features of the system (for example, (12) denotes the permutation of H 1 and H 2 of water) as illustrated in Fig. 1.1. The resultant set of permutations now takes on a coldly abstract form which is ideally suited to manipulation using the analytical methods associated with the theory of groups of equivalent particles. The theorems necessary for this analytical process will be introduced later with reference to relevant detailed texts but it is not intended to present comprehensive proofs.

1.2. The Role of Symmetry in Spectroscopy

The symmetry possessed by a molecule may be used to advantage in simplifying the calculation of the energy levels of the system and deciding which transitions are or are not allowed between them in all fields of spectroscopy. It is of some importance to realise that the observed structure and effective symmetry of a molecule depends upon the technique chosen to study it. The time scale of ultraviolet spectroscopy is very short and it is possible to study very short-lived species which may include different conformers or rotamers each with its own symmetry elements. Infrared spectroscopy does not have as short a time-scale but it can still be used to 'freeze' and study conformers or rotamers of a given molecule at appropriate temperatures. The molecular model at radiofrequencies, where the time scale is long compared with the internal vibrations and/or rotations, is very often the configuration averaged over these rapid motions, except perhaps at very low temperatures.

1.3. Symmetry in Nuclear Magnetic Resonance

The symmetry of a molecule relevant to the nuclear magnetic resonance (NMR) problem may be quite different from the obvious symmetry of the molecule as a whole. This may arise because a limited number of nuclei in the molecule have effective nuclear magnetic moments and/or because of 'rapid' internal motions and/or because some of the obvious symmetry elements have no relevance in the NMR problem or because of quantum mechanical reasons. The symmetry expressed as either symmetry elements or permutations may be less than, or greater than, the symmetry of the molecule as a whole [1]. Examples are given later.

The symmetry of an NMR system is reflected in the equality of coupling constants characterising the interaction between nuclei and the equality of chemical shifts of nuclei. The theoretical treatment and its applications are founded upon these equalities.

There are two symmetry types which will be considered here. The first comes under the heading "rigid molecules" which excludes molecules involving fast reversible equilibria between intramolecular conformers such as ethane or 1-phenyl, 2-bromo ethane. The meaning of 'fast' in this context is defined as indicating that the residence of times of a molecule in its possible conformers are very much less than the reciprocal of the appropriate chemical shift difference between nuclei rendered non-equivalent in the 'static' conformer or the appropriate coupling constant expressed in sec^{-1} units (Hz) [2]. The molecular configuration (or nuclear configuration) of rigid molecules used to define the symmetry is that averaged over internal vibrations. Benzene derivatives with single nucleus or linear substituents are regarded as rigid, for example.

The second symmetry type may be less familiar, but is no less important. This includes 'non-rigid' molecules where there are internal rotations which are 'fast' on the NMR time scale such as propane or fast conversion between conformers as in cyclohexane at normal temperatures. The second classification also includes molecules such as PF_5 where scrambling of the fluorine nuclei takes place and the so-called ring 'whizzer' compounds. Examples are given later.

It has already been stated that the main purpose of this paper is to describe the methods by which it is possible to extract the information which symmetry will provide. However it would inhibit the understanding of the usefulness of symmetry in NMR if the alternative methods of simplifying the problems were ignored. The alternative methods specifically include the use of magnetic equivalence and the 'X approximation'. These are therefore considered in some detail and in particular the relationship between symmetry and magnetic equivalence is discussed in sections 3 and 4.

The discussion has so far been very qualitative and it is necessary to consider the symmetry properties of molecules in more quantitative detail, related to group theory, to facilitate an understanding of the examples which will be given later.

2. Symmetry and Symmetry Groups of Rigid and Non-rigid Systems

The symmetry of a 'rigid' system is easily defined and assimilated in geometric terms but in the case of non-rigid molecules it is not possible to define 'symmetry' in the same visual terms. It is possible to define the *symmetry groups* of both rigid and non-rigid molecules however in similar terms which are, of necessity, more abstract involving permutations of identical particles (nuclei in this case but may include electrons in the more general group) which constitute the system. The idea of the symmetry group can be introduced most easily using the visual geometric features of rigid systems. It is necessary in the first instance to distinguish between *symmetry elements* and *symmetry operations*.

2.1. Symmetry Elements

A symmetry element is a geometrical feature such as a line (axis of symmetry C), a plane (σ) or a point (centre of symmetry, i). These are illustrated in the example of cyclobutane Fig. 2.1 which also allows us to distinguish between a proper axis C and an improper axis S of symmetry.

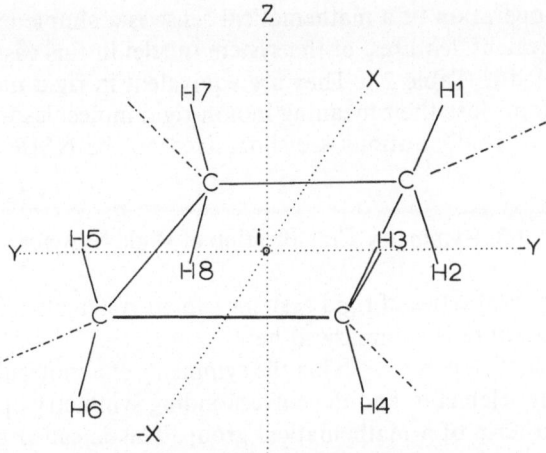

Fig. 2.1. Symmetry elements and symmetry operations for cyclobutane with D_{4_h} symmetry

2.2. Symmetry Operations

A symmetry operation in the geometric sense is a systematic change of the chosen co-ordinates of each point in a system such that the end-product cannot be distinguished from the original configuration. This may involve rotation of the system as a whole (C), possibly followed by a reflection (S), a reflection alone (σ) or an inversion through a centre of symmetry (i).

Table 2.1. *Symmetry elements and the corresponding symmetry operations of cyclobutane relevant to the NMR problem*

Symmetry Element		Symmetry Operation	
Geometrical Feature[a]	Symbol	Description	Permutation[a]
Z proper axis	C_4^+	Rotation clockwise by $2\pi/4$	(1357)(2468)
	C_4^-	Rotation anticlockwise by $2\pi/4$	(1753)(2864)
Z proper axis	C_2	Rotation by $2\pi/2$	(15)(37)(26)(48)
X proper axis	C_2	Rotation by $2\pi/2$	(18)(27)(45)(36)
Y proper axis	C_2	Rotation by $2\pi/2$	(14)(23)(67)(58)
X = Y proper axis	C_2	Rotation by $2\pi/2$	(34)(78)(16)(25)
X = −Y proper axis	C_2	Rotation by $2\pi/2$	(12)(56)(47)(38)
X = Y = Z = 0 point	i	Inversion through centre	(16)(25)(47)(38)
Z improper axis	S_4^+	Rotation clockwise by $2\pi/4$ then reflection in the XY plane	(1458)(2367)
	S_4^-	Rotation anticlockwise by $2\pi/4$ then reflection in YX plane	(1854)(2763)
XY plane	σ_h	Reflection	(12)(34)(56)(78)
XZ	σ_v	Reflection	(17)(28)(35)(46)
YZ plane	σ_v'	Reflection	(13)(24)(57)(68)
(X = −Y), Z plane	σ_d	Reflection	(37)(48)
(X = Y), Z plane	σ_d'	Reflection	(15)(26)

[a] Appropriate for Fig. 2.1.

A symmetry operation in a mathematical sense is a simple exchange or permutation of equivalent 'features' of the system (nuclei in this case). These definitions are illustrated in Table 2.1. They are equivalent in rigid molecules but the geometric definitions lose their meaning in non-rigid molecules where the molecular conformers or configurations are short-lived on the NMR time-scale.

2.3. Symmetry Classification of Rigid Systems

The symmetry properties of rigid systems can be systematically classified and the necessary procedure is summarised here.

The essential first step in classifying the symmetry of a molecule is to establish the set of symmetry elements, and the corresponding symmetry operations, which satisfy the four criteria of a mathematical group. This is called the *complete set* of elements or operations. The criteria of a mathematical group are set out in more detail in appendix B and summarised here.

The first criterion requires that all the 'products' of the operations taken two at a time is also an operation in the set. The term 'product' here is taken to mean the successive application of the operations.

The second requirement defines an operation which does nothing at all to the system. This is the identity operation E, and the result of successive application of E and any other operation X, independent of the order of application, is the same as the application of the operation X above, $XE \equiv EX \equiv X$.

The associative law of combination of symmetry operations must be valid, $X(YZ) = (XY)Z$.

Finally there must exist for each operation X a second operation X^{-1} which exactly reverses the effect of the first so that $XX^{-1} = X^{-1}X = E$. X^{-1} is called the inverse of X.

Once the complete set has been established the appropriate symmetry group can be determined by reference to long standing tables or logically deduced using a systematic method [3].

Systematic Method of Symmetry Classification

It is often expedient to look first for the special groups characteristic of linear molecules and highly symmetric molecules.

Linear Molecules

Linear molecules which do not possess a centre of symmetry belong to the $C_{\infty v}$ classification (E, $2C_{\infty}^{\varphi}$, $\infty \sigma_v$). Those with a centre of symmetry are labelled $D_{\infty h}$ (E, $2C_{\infty}^{\varphi}$, $\infty \sigma_v$ i $2S_{\infty}^{\varphi}$, $\infty C2$).

Highly Symmetrical Molecules — the Cubic Groups

Molecules belonging to the tetrahedral group T_d (E, $8C_3$, $3C_2$, $6S_4$, $6\sigma_d$), the octahedral group

$$O_h (E, 8C_3, 6C_2, 6C_4, 3C_2 (= C_4^2) i 6S_4, 8S_6, 3\sigma_h, 6\sigma_d)$$

and Icosahedral group I_h (see reference *3*) are normally obvious because of their high symmetry.

Other Molecules

It is necessary to begin a systematic search for symmetry elements if none of the above are appropriate. The systematic method is illustrated in a schematic way in Table 2.2.

Table 2.2. *Systematic symmetry classification of molecules (not belonging to special classifications)*

The first element which is sought is a proper axis of symmetry C_n

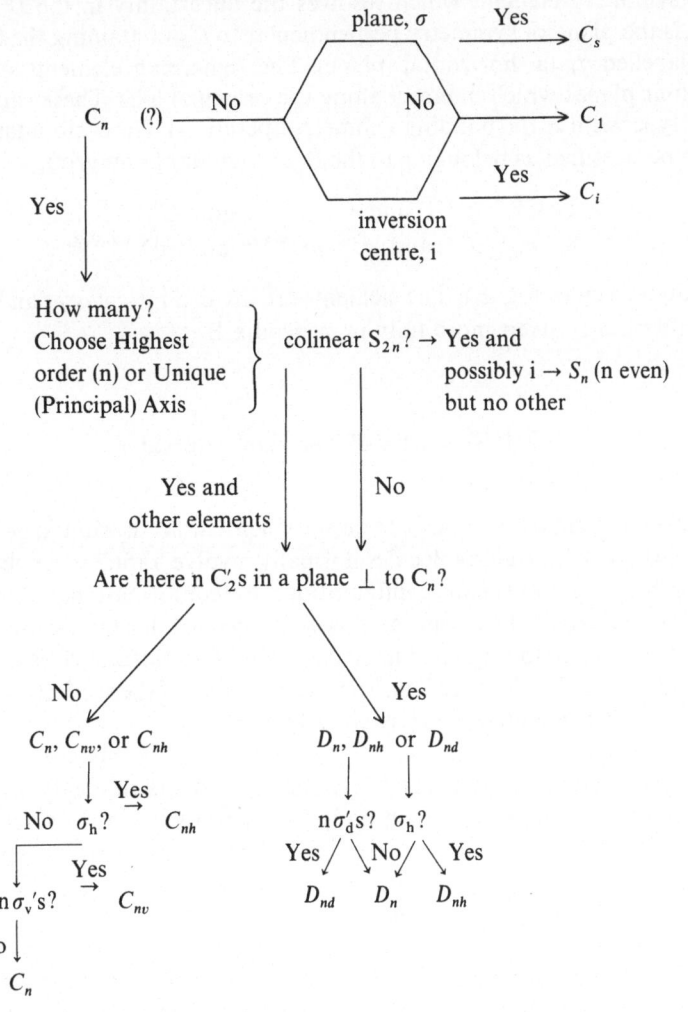

Example of Symmetry Classification: Cyclobutane

This molecule has been used to illustrate the types of symmetry elements which molecules may passes. In this example it is assumed that the time-averaged planar carbon skeleton represents the equilibrium configuration of the molecule. The scheme in Table 2.2 can be used as follows.

Cyclobutane has one C_4 (the principal) axis and five C_2 axes one of which is coincident with the principal axis. These features immediately rule out C_s, C_i and C_1.

There is an S_4 improper axis colinear with the principal axis. There is a centre of symmetry and it is possible to pick out a number of planes of symmetry. The molecule does not therefore belong to the S_n (n-even) classification.

There are four C_2 axes in the plane of the carbon skeleton ring perpendicular to the principal axis and this means that it is necessary to consider the D_n, D_{nh} and D_{nd} groups, rejecting the C_n, C_{nv} and C_{nh} groups.

The symmetry element which resolves the uncertainty in the D type classification is the plane of symmetry perpendicular to C_4 containing the four carbon atoms, labelled σ_h (a 'horizontal' plane). The remaining elements of symmetry are the four planes which intersect along the principal axis. These can be divided into two types with corresponding symmetry operations. The cyclobutane molecule can then be classified as belonging to the D_{4h} symmetry point group with elements

$$E, C_4^Z, C_2^Z, C_2^Y \text{ and } C_2^X, C_2^{X=Y} \text{ and } C_2^{X=-Y}, i$$
$$S_4, \sigma_h^{XY}, \sigma^{XZ} \text{ and } \sigma^{YZ}, \sigma^{(X=Y)Z} \text{ and } \sigma^{(X=-Y)Z}$$

which can be seen in Fig. 2.1. The elements are divided into *classes* and the reason for this division is given more fully in appendix B.

2.4. Symmetry of Non-rigid Molecules

It is of fundamental importance to decide when a molecule should be considered to be 'non-rigid'. Non rigid molecules normally involve 'rapid' reversible dynamic equilibria between molecular configurations or conformations which may or may not be equivalent. The term rapid is a relative one depending upon the technique used to observe the equilibrium and also the conditions, such as temperature, under which the molecule is studied, Specific examples of these effects will be given when the NMR symmetry group is considered.

It is not obvious from the time dependent nature of the configurations of non-rigid molecules if the concept of symmetry elements so ideally suited to the description of the symmetry of rigid molecules has any meaning when discussing the symmetry of non-rigid molecules. Furthermore it is essential that any new definition of a symmetry group which can be applied to non-rigid molecules should also apply to rigid molecules.

The theory was first expanded to include non-rigid molecules by LONGUET-HIGGINS [4] actively supported by HOUGEN [5]. The basis of the new definition of a symmetry group is the permutation of identical particles (to be compared with the symmetry operations of rigid groups) and the introduction of a further operation E* which is the inversion of all particle positions through the centre of mass.

The definition of the more general symmetry group as given by LONGUET-HIGGINS is: —

E   the identity operation

P   any permutation of positions and spins of identical nuclei or any product of such permutations.

E* inversion of all particle positions through the centre of mass.

P* permutation inversions = PE* = E*P.

It is not necessary to include all of the elements of the most general group for most molecules. This is because the timescale of a given laboratory experiment may be too short to allow certain nuclear permutations ever to occur. The molecular symmetry group is then composed of feasible elements only. It includes: —

1. all feasible P including E

2. all feasible P* not necessarily including E*.

The decision as to whether or not a permutation is feasible must be taken 'in situ' with the conditions of the experiment in mind. Ethane, for example, is a gas at normal temperatures with a low barrier to internal rotation. It is possible therefore to test the feasibility of all conceivable permutations using the notation in Fig. 2.2.

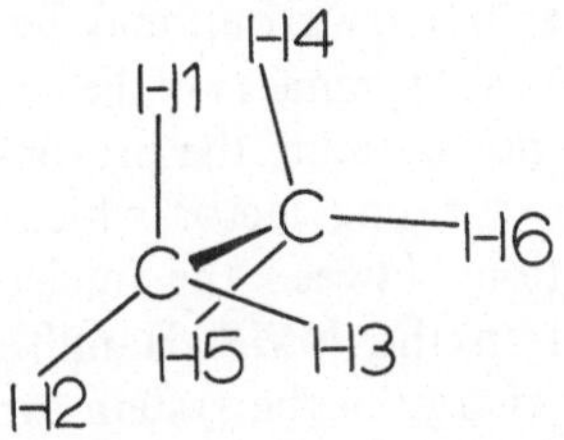

Fig. 2.2. Model for non-rigid ethane

The permutations (12) and (12) (45), for example are not feasible because carbon-hydrogen bonds must be broken in order to superimpose the resultant configurations on the original. On the other hand (123) and (12) (45)* are feasible. The result of the permutation (123) can be counteracted by an intramolecular rotation which is feasible in this context. However it must be emphasised that the permutation itself is not to be regarded as a physical motion in any sense. The permutation inversion (12) (45)* involves two basic steps. The first is a permutation of the hydrogen atoms 1 and 2 accompanied by the permutation of 3 and 4 simultaneously. This changes the 'sense' of the two methyl groups but the second operation, an inversion through the centre of mass, reverses the sense of both methyl groups again. The resultant configuration can be superimposed on the original by simple rotations of the molecule as a whole in space. It is necessary to extend the idea of a permutation of the group leaving the molecule in a configuration indistinguishable from the original and add that it may be necessary to view the molecule from a different direction or move it as a whole in space in the way described above.

More recently ALTMANN has proposed an alternative definition of the symmetry group of non-rigid molecules [6].

The two approaches given by LONGUET-HIGGINS and ALTMANN differ fundamentally. The geometrical structure of a non-rigid molecule may be thought to change continuously and LONGUET-HIGGINS has chosen that structure in which the parameter describing the motion (torsion eg) has an arbitrary value. ALTMANN however has shown that important information can be obtained from the mathematical groups of all structures that correspond to extremes of energy since such structures must contain some elements of symmetry.

The LONGUET-HIGGINS approach will be used here since it is believed to be conceptually simpler but no claim is made that it is mathematically superior to the ALTMANN approach. It may have been noted that no great detail has been provided even for the LONGUET-HIGGINS definition of the group, in terms of examples. There are good reasons for this, as will be seen later, since the NMR symmetry group involves a further modification of the LONGUET-HIGGINS general group.

3. The NMR Permutation Group and Symmetry

It has already been intimated in the introduction that the NMR symmetry group is in general different from the molecular symmetry group. One factor which influences the NMR symmetry group may be considered to be redundant geometrical features. The planes of symmetry of the benzene and ethylene molecules containing all the nuclei do not permute the protons in either molecule and are therefore redundant. The over-riding factor which influences the NMR symmetry group is the interaction between the nuclei of the system. The NMR symmetry group is implicit in the NMR Hamiltonian which is a quantum mechanical summary of the energy of the system expressed in terms of the ZEEMANN energy (interaction of the nuclear magnetic dipoles with the external magnetic field) and the interaction between nuclear magnetic dipoles which may be direct or indirect. This follows quite simply from the equality of coupling constants and chemical shifts of the system of nuclei. The interactions which are considered important therefore determine exclusively the NMR symmetry group.

3.1. Symmetry and the NMR Hamiltonian

The form which the NMR Hamiltonian takes is so important in deciding the appropriate symmetry group that the dominant interactions are summarised here in Hamiltonian form —

Zeeman Interaction, H°

$$H^0 = \sum_i v_i I_{zi}$$

v_i Larmor precessional frequency for nucleus: (sec^{-1})

I_{zi} z component of spin angular momentum for nucleus i.

$\sum_i$ sum over all nuclei in the system.

Scalar (Indirect, Electron-coupled) Dipole-Dipole Interaction, H^J

$$H^J = \sum_{i<j} \sum J_{ij} I_i I_j$$

J_{ij} spin-spin coupling constant between nuclei i and j

I_i spin angular momentum operator for nucleus i.

$\sum_{i<j} \sum$ summation over pairs of nuclei ij such that the interactions are included once only in the Hamiltonian between a given pair.

Direct Dipole-Dipole Interaction, H^D

$$H^D = \sum_{i<j} \sum D_{ij} \left[I_{zi} \cdot I_{zj} - 0.5 \left[I_{xi} \cdot I_{xj} + I_{yi} I_{yj} \right] \right]$$

D_{ij} direct dipole-dipole coupling constant.

I_{xi}, I_{yi} x and y components of spin angular momentum for nucleus i.

Two other contributions, quadrupole coupling interactions for spins $> 1/2$ and anisotropic electron-coupled dipole-dipole interactions, should be mentioned but since they do not add anything new to the symmetry problem they will not be considered in detail.

The NMR symmetry group is most satisfactorily defined as containing all the nuclear permutations which do not alter the nuclear parameters [1]. These include the identity E and the appropriate permutations P of the LONGUET-HIGGINS group. The permutation inversion P* permute the NMR parameters in the same way as P since an inversion through the centre of mass does not permute the nuclei of the molecule.

It is now possible to consider in more detail general and specific examples of the influence of the Hamiltonian on the NMR symmetry group. The discussion here is concerned with systems composed of spin 1/2 nuclei but where relevant, differences arising when the spin quantum number $>1/2$ will be discussed.

3.2. Isolated Groups of Nuclei (Same Isotopic Species)

The chemical shifts of the nuclei in isolated groups are the same i.e. the group is composed of *isochronous* nuclei. These groups can be sub-divided into two distinct types.

a) All the J coupling constants between the nuclei within the group are identical. It normally follows that all the D coupling constants will have identical values as well ($\neq$ J!). The nuclei are then *magnetically fully equivalent* [1] [7,8] as in the methane C-12.

b) There is more than one unique J coupling constant characterising the interaction between the nuclei of the group. It follows that there will be a corresponding number of D coupling constants. This is the case in ethylene C_2-12.

---

[1] The terminology of magnetic full equivalence, magnetic equivalence and magnetic non-equivalence is introduced here in a qualitative way. A more quantitative definition will be given later in discussing the factorisation of the secular determinant of the Hamiltonian matrix.

a.1. The Example of Methane in an Isotropic Medium

The Hamiltonian for methane in an isotropic medium can be summarised as

$$H = H^0 + H^J.$$

H^D depends upon the relative orientations of the magnetic dipoles and averages to zero because of the random motions of molecules in the isotropic medium (gas or mobile liquid). It can be shown from quantum mechanical arguments [9] that the spectrum is independent of the scalar interactions between magnetically equivalent nuclei H^J (provided the interest lies in the transition energies alone and not in the NMR relaxation processes associated with the system). It is therefore possible simply for the purpose of illustrating the nature of the symmetry group in this case to put all the J coupling constants equal to zero. The operations of the symmetry group are then the factorial n, ($= 24$ in this case) permutations of the isochronous nuclei. Representative examples of the symmetry operations are shown in Table 3.1.

Table 3.1. *Symmetry operations of the permutation group for 4 nuclei*

1 E	identity
6 (12)	two at a time
3 (12) (34)	two pairs at a time
8 (123)	three at a time
6 (1234)	four at a time

The total number of symmetry operations is called the *order* of the group, h.

a.2. The Example of Methane in an Anisotropic Medium

Quantum mechanics now indicates that the $J_{ij}I_iI_j$ and $D_{ij}I_iI_j$ terms may be dropped from the Hamiltonian leaving v_iI_{zi} and $D_{ij}I_{zi}I_{zj}$ terms only. The full permutation group has an order of 24 as above.

a.3. Groups of Nuclei with Spin $\geqslant 1$

The symmetry group is the full permutation group for the n nuclei when the molecule containing the spin $\geqslant 1$ nuclei is in an isotropic medium since the scalar interaction may be ignored and the direct interaction averages to zero.

However, when the molecule is in an anisotropic medium where the direct interactions do not vanish the additional terms in the Hamiltonian result in each term, including the scalar interaction, playing a full part in determining the energies of the eigenstates of the molecules. Furthermore these contributions cannot be disregarded in the way applicable to spin 1/2 systems and both D and J appear directly or indirectly in the spectrum. A further consequence is that although the nuclei of a group with spin $\geqslant 1$ may be magnetically equivalent in the isotropic phase they are magnetically nonequivalent in the anisotropic medium [8, 10].

b.1. The Example of Ethylene in an Isotropic Medium

The Hamiltonian consists of $\nu_i I_{zi}$ and $J_{ij} I_i I_j$ terms only for this example and it is possible to ignore $J_{ij} I_i I_j$ as in a. 1. above. Once again the appropriate symmetry group has an order of 24 but now the protons of ethylene are magnetically equivalent not magnetically fully equivalent.

b.2. The Example of Ethylene in an Anisotropic Medium

There are three unique J and D coupling constants in ethylene (gem, cis and trans). The nuclei can therefore no longer be magnetically equivalent since quantum mechanics indicates that for this case the full Hamiltonian $H^0 + H^J + H^D$ must be used. The NMR symmetry group is composed of nuclear permutations which exchange chemical shifts (all equal) geminal couplings, cis couplings and trans couplings. The group can be summarised, if the nuclei are labelled 1, 2, 3 and 4, as

$$E, (13)(24), (12)(34), (14)(23)$$

This group, of order four, is *isomorphic* with the symmetry point groups D_2 and C_{2v}. Two groups G and G' are said to be isomorphic if to each operation $P_1 P_2 P_3 \ldots$, of G there corresponds an element $P'_1 P'_2 P'_3 \ldots$ of G's so that if $P_1 P_2 = P_3$ then also $P'_1 P'_2 = P'_3$.

This symmetry group is also of importance in those cases where the four spin 1/2 nuclei interact with nuclei outside the group such that there is more than one unique coupling constant between the groups. An example will be considered later.

3.3. Interacting Groups of Spin 1/2 Nuclei

It is again necessary to distinguish between magnetic full equivalence, magnetic equivalence and magnetic non-equivalence of the nuclei of the interacting groups. The additional terms which must be added to the sum of the Hamiltonians of the interacting groups are those involving the interactions between the groups. There are two distinct cases for two interacting groups where (i) there is one unique coupling constant only between the nuclei of the one group and every other nucleus and (ii) where there is more than one unique coupling constant between the groups. It will become apparent that combinations of these are often encountered where the system is composed of more than two groups [11].

The example of interacting magnetically fully equivalent groups. Notation. The notation used throughout this paper follows the original papers whereever possible.

The System $A_m B_n C_p \ldots$

The nuclei of the A group (and similarly the other groups) are magnetically fully equivalent because they are isochronous and 1) the A nuclei are equally coupled to one another J_{AA}, D_{AA}. 2) the A nuclei are equally coupled to each nucleus of the other groups by J_{AB}, J_{AC}... and D_{AB}, D_{AC}.

The notation also implies that the groups of nuclei are strongly coupled i.e. $|J_{AB}^2/(v_A - v_B)| > \Delta$ where Δ is the linewidth (resolution) or precision of the parameters required.

The same notation is used for anisotropic media when $|D_{AB}^2/(v_A - v_B)| > \Delta$. The straight brackets mean take the modulus or simply the magnitude without regard for sign.

The 'symmetry' group of each of the groups of magnetically fully equivalent nuclei is the appropriate permutation sub-group P_m, P_n, P_p and the 'symmetry' group for the whole system is the combination of the individual permutation sub-groups $P_m x P_n x P_p \ldots x \ldots$. This is valid for both isotropic and anisotropic phase spectra. Ethyl iodide $CH_3 \cdot CH_2 \cdot I$, has been given as an example of an $A_3 B_2$ system [8]. It is relevant to note that this and similar systems, propanes, butanes, etc., can be considered as non-rigid systems under normal conditions.

The Example of Interaction between Magnetically Equivalent and Magnetically Non-equivalent Groups
$AA'X_n X_n'$ and $AA'B_n B_n'$ Systems

Systems of these types have been discussed at length with reference to the isotropic phase spectra where the symmetry group is the appropriate combination of the permutation sub-groups P_n and P_n' and also the C_2 symmetry sub-group which arises from the geometrical element relating the primed and non-primed parts of the systems. The systems can be written alternatively as $X_n \cdot A \cdot A' \cdot X_n'$ and $B_n \cdot A \cdot A' \cdot B_n'$ to illustrate this symmetry property. The notation now distinguishes between magnetically non-equivalent groups by the primes and the A,X terminology infers that $|J_{AX}^2/(v_A - v_X)| < \Delta$.

A specific example of an $AA'X_3 X_3'$ system is fluoro-N,N'-dimethyl-1,3,2,4-diazadiphosphetidine [11] where the phosphorus and fluorine nuclei constitute the system of interest.

$$
\begin{array}{c}
PF_3 \\
\diagup \quad \diagdown \\
CH_3 - N \qquad N - CH_3 \\
\diagdown \quad \diagup \\
PF_3
\end{array}
$$

Symmetrically 2,3 disubstituted butanes

$$CH_3 \cdot CHX \cdot CHX \cdot CH_3$$

may be classified either as $AA'X_3 X_3'$ or $AA'B_3 B_3'$ depending upon the substituent X and the operating conditions – magnetic field etc.

The 'symmetry' group which will henceforth be referred to as the permutation group for such systems has an order of seventy-two since there are that many permutation operations in the group. The group is summarised in Table 3.2, where the X or B nuclei have been labelled 1 to 6, 123 for X_3, B_3 and 456 for X'_3, B'_3, and the A nuclei labelled 7 and 8 [1].

Table 3.2. *Permutation operations allowed for the $AA'X_3X'_3$ and $AA'B_3B'_3$ systems*

E	12 (123) (45)
4 (123)	6 (12)
4 (123) (456)	6 (14) (25) (36) (78)
9 (12) (45)	12 (142536) (78)
	18 (1425) (36) (78)

E is the identity operation (see appendix B) and the pre-fixed numbers indicate the number of symmetry operations in a given class. It is worth emphasising here that combinations of permutation operation of odd (123) and even (45) parity are allowed because they leave the NMR Hamiltonian unchanged. There are twelve operations in the class represented by (123)(45) and four in the class represented by (123).

The permutation group for the $AA'X_nX'_n$ and $AA'B_nB'_n$ systems in anisotropic media is the same as in isotropic media when the nuclei within the X_n and X'_n are respectively magnetically fully equivalent.

A Second Example of Interaction between Magnetically Equivalent and Non-equivalent Groups of Nuclei
The $A_2A'_2BB'B''B'''$ and $A_2A'_2XX'X''X'''$ Systems

The α, α' disubstituted p-xylene molecules may belong to this classification if conditions are such that it may be assumed that there is a rapid dynamic equilibrium between conformers of the molecules involving rotation about the terminal carbon-carbon bonds. The operations of the permutation group are given in Table 3.3 where the numbering notation corresponds to the labelling in Fig. 3.1.

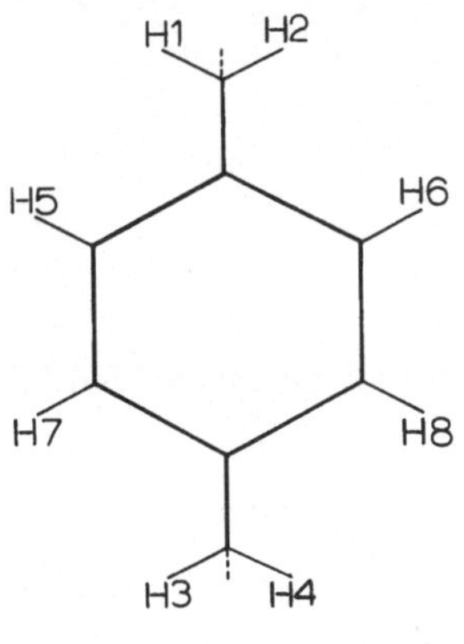

Fig. 3.1. Model for the non-rigid $A_2A'_2BB'B''B'''$ system — α,α' disubstituted p-xylene

Table 3.3. *Permutations operations consti-*
tuting the group for $A_2A_2'BB'B''B'''$ *systems*

E	2 (12)(56)(78)
(12)(34)	2 (13)(24)(58)(67)
(56)(78)	2 (13)(24)(57)(68)
(12)(34)(56)(78)	2 (1324)(57)(68)
2 (12)	2 (1324)(58)(67)

The total order of the group is sixteen and it will be shown in appendix B that this group is isomorphic with the familiar D_{4h} symmetry group. This system has been chosen to illustrate the determination of the permutation group, the derivation of the character table, the classification and construction of symmetrised wavefunctions. The details are given in appendix B.

Interacting Groups of Magnetically Non-equivalent Nuclei AA′BB′ and AA′XX′ Systems

The symmetry groups appropriate for these systems are $C_s(E,\sigma)$, $C_2(E,C_2)$ and $C_i(E,i)$, examples are given in Fig. 3.2(a) [12, 13].

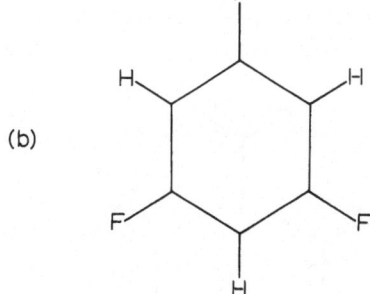

Fig. 3.2. (a) Models for the rigid AA′BB′ system. (b) Model for the rigid AA′A″BB′B″ system

AA′A″BB′B″ and AA′A″XX′X″ Systems

These systems are characterised by one Hamiltonian and therefore by one permutation group which is isomorphic with D_3 or D_{3r}. Examples are given in Table 3.2(b) [14].

AA′A″A‴BB′B″B‴ and AA′A″A‴XX′X″X‴ Systems

The eight spin system of this type are rather more interesting from the symmetry point of view than the two previous examples because of the diversity of possibly permutation groups [15]. These are illustrated in Fig. 3.3 where the Hamiltonian is summarised in terms of the unique coupling constants within and between the two groups below each diagram. The isomorphic symmetry groups are used to characterise the systems.

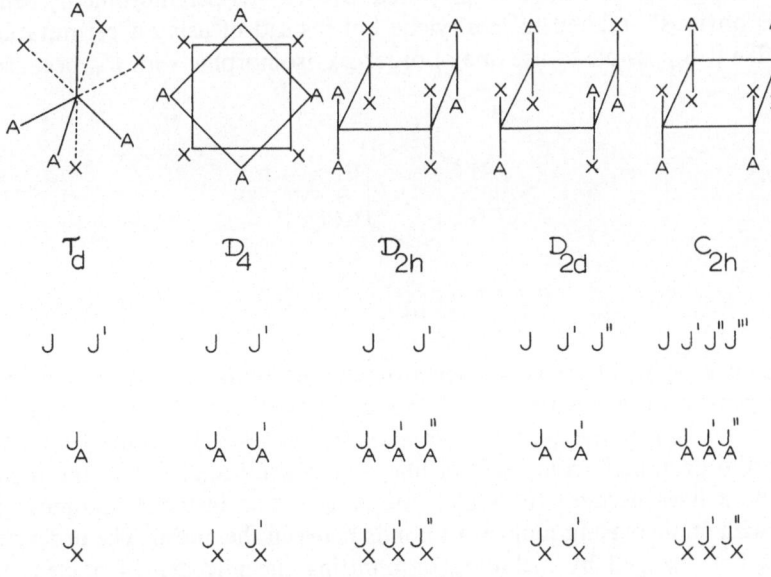

Fig. 3.3. Models for AA′A″A‴BB′B″B‴ systems with T_d, D_4, D_{2h}, D_{2_d} and C_{2_h} symmetry characterised by the unique scalar coupling constants

AA′A″$B_n B_n′ B_n″$ and AA′A″$X_n X_n′ X_n″$ Systems

The first member of these systems can be identified with the systems in Fig. 3.2(b). The examples with n = 2 and 3 are manifested in the α,α′,α″-trisubstituted mesitylenes and mesitylene itself. The order of the group for the trisubstituted mesitylene is 48 [16] and for mesitylene-type systems, it is 1296 [17]. The latter is a very unwieldy permutation group indeed but will be shown later that in this and many other of the examples considered it is not necessary to resort entirely to group theoretical methods in working with such systems.

Systems Containing Spin ≥ 1 Nuclei

Nuclei with spin quantum numbers I greater than one half have electric quadrupole moments as well as nuclear magnetic dipole moments. In general, where the nuclei are in unsymmetrical environments these electric quadrupole moments interact with oscillating electric fields giving rise to a very efficient relaxation mechanism which reduces the spin lattice relaxation time of the nucleus. The interaction with adjacent spin 1/2 nuclei is therefore very often averaged to zero because while the spin 1/2 nucleus remains in a given orientation with respect to the magnetic field the nucleus with spin $> 1/2$ changes its orientation many times. The result may be (i) zero coupling between spins with spin $> 1/2$ and spins equal to 1/2 (ii) non zero, but time-averaged coupling which leads to broad lines (iii) non-zero-couplings. The Hamiltonian must of course be changed accordingly but the inclusion of the additional nuclei may or may not alter the permutation group.

An example of (i) above is provided by 1,3,5-tri-perchloromethylbenzene which is obviously related to mesitylene but instead of using a permutation of order 1296 it is possible to use one of order six, isomorphic with D_3, because the

$$D_3$$

chlorine nuclei spin 3/2 are very efficiently relaxed. Pyridine is an example where the nitrogen-14 nucleus with spin 1 is not completely decoupled from the ortho protons of the ring by the quadrupolar relaxation process. The multiplet assigned to the ortho protons is composed of lines which are broader than the meta or para proton lines because the scaler coupling in this instance becomes much smaller with an increasing number of bonds between the nuclei. The order of the group is not changed by including or omitting the nitrogen-14 nucleus from the problem and the permutation group is isomorphic with C_2 or C_s.

$$C_s \text{ or } C_2$$

A further example here of 1,3,5-triperdeuteromethylbenzene illustrates the change in the permutation group according to which interactions are considered as being important.

The scalar coupling between the protons and deuterons will be very small in this molecule since in mesitylene itself the methyl proton ortho ring proton interaction is characterised by $J = 0.89 \pm 0.10$ Hz. Therefore in isotropic phase spectra where the resolution of the instrument may be 0.3 Hz or slightly better the spectrum of the protons will appear as a slightly broadened single line. An approximation can therefore be made isolating the protons and deuterons separately with the appropriate permutation groups, isomorphic with D_3 and C_{3_v} respectively. This exercise may appear wholly academic but when the molecule is included in an anisotropic medium where the dipole-dipole interaction cannot be considered to average to zero, it becomes apparent that now since the dipole-dipole interaction in such systems is certainly not small the full permutation group must be employed i.e.

$$D_3 \times C_{3_v} \times C_{3_v} \times C_{3_v}$$

as in mesitylene itself.

The Effect of Time-averaging Processes on the Hamiltonian and Permutation Group

It is implicit in many of the systems which have been discussed that dynamic equilibria can lead to an enlarged permutation group for the system. The molecule 1,chloro-2,iodoethane at high temperatures will have an effective permutation group of order four while it is conceivable that at low enough temperatures the molecule will be restricted in its ability to rotate about the carbon-carbon bond giving an effective group of order two for the *trans* conformer. In the first case the nuclei of the two groups are magnetically equivalent A_2B_2, in the second, magnetically non-equivalent AA'BB'.

Other forms of time averaging process are possible and a particularly interesting one manifests itself in fluxional molecules such as π-cyclopentadienyl iron dicarbonylcyclopentadienyl shown in Fig. 3.4 [18].

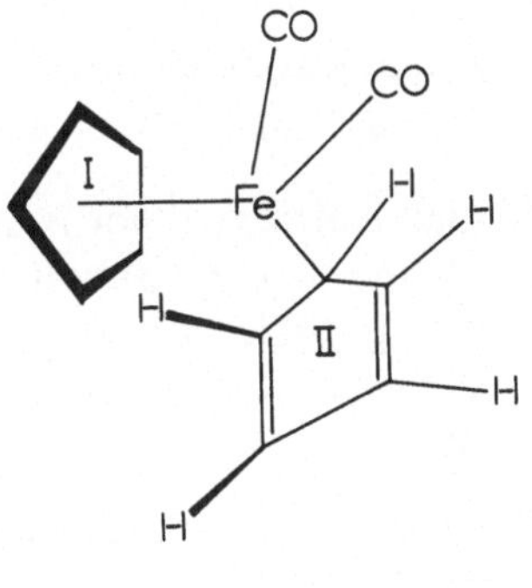

Fig. 3.4. The π-cyclopentadienyliron dicarbonylcyclopentadienyl molecule — an example of a 'fluxional' molecule. The cyclopentadienyl ring, II, has a temperature-dependent 'local' symmetry

The proton spectrum of this molecule at high temperatures (120° C) consists of two lines which may be assigned to the protons of the π-bonded cyclopentadienyl ring (I) and the protons of the σ-bonded cyclopentadienyl ring (II). The separate single lines at this temperature indicate that both rings are engaged in dynamic 'motions' which average the environments of the protons in each ring distinctly as indicated by the different chemical shifts. The separation between the rings is large and therefore interaction between the protons of the two rings can be neglected. The appropriate permutation group is therefore P_5 for both rings having an order of 120.

The spectrum changes in appearance as the temperature is lowered. The line at low field first broadens and then splits leaving the high field line unaltered. The spectrum reaches a maximum multiplicity at about $-80°C$ and does not change in appearance down to $-100°C$. The π-bonded cyclopentadienyl ring can be assumed to be 'rotating' freely even at such temperatures and the permutation group for that ring system remains P_5. The dynamic motion of the σ-bonded ring has obviously been reduced to a level where the ring can be thought of as conventionally bonded to the iron atom. The protons of the σ-bonded ring then belong to the AA'BB'X classification and the N.M.R. permutation group is isomorphic with C_s or C_2.

4. Factorisation of the Determinant of the Hamiltonian Matrix

The steps involved in calculating NMR spectra are as follows [12]:

1. decide on the Hamiltonian H,
2. choose a set of wavefunctions ψ_n,
3. derive the Hamiltonian matrix elements $(\psi_n H \psi_n)$,
4. diagonalise the determinant of the Hamiltonian matrix to give the energy levels, E_n,
5. calculate the stationary state wavefunctions,
6. calculate the transition energies and intensities.

The order of the Hamiltonian matrix is 2^N for N spin $1/2$ nuclei and it can be seen that even for three nuclei the diagonalisation of Hamiltonian matrix becomes exceedingly difficult. It is possible however to factorise the determinant of the Hamiltonian matrix, reducing it to the product of a number of independent matrices of lower order, by using

1. total magnetic quantum number $m_T = \sum_i m_i$,
2. total spin quantum number I_T,
3. symmetry factorisation.

4.1. Total Magnetic Quantum Number Factorisation

Every wavefunction for the system may be characterised by a total magnetic quantum number m_T = sum of magnetic quantum numbers for all the nuclei $\sum_i m_i$. The magnetic quantum number for a given nucleus is simply the 'expectation' or eigenvalue of the z component of spin angular momentum and for spin 1/2 nuclei this is $+1/2$ or $-1/2$ depending upon whether the spin is 'parallel' or 'anti parallel' respectively to the external magnetic field. The $+1/2$ state is characterised by spin wavefunction α and the $-1/2$ state by β. In systems containing more than one nucleus the spin states are characterised by basic spin product wavefunctions $\alpha\beta\alpha\alpha\ldots$, for example, and each of these is characterised by an m_T value; for $\alpha\beta\alpha\alpha$. $m_T = +1/2 - 1/2 + 1/2 + 1/2 = +1$. The Zeeman term, abbreviated as $\sum_i v_i I_{zi}$, dominates the Hamiltonian matrix elements so that off-diagonal elements involving inter-nuclear interactions (J and D) only may be neglected [12]. This leads to a binominal factorisation of the Hamiltonian matrix for systems composed of spin 1/2 nuclei and this is illustrated in Table 4.1.

Table 4.1. *Classification of wavefunctions according to m_T for up to 5 spin 1/2 nuclei*

m_T	N = 2	3	4	5
5/2				1
2			1	
3/2		1		5
1	1		4	
1/2		3		10
0	2		6	
-1/2		3		10
-1	1		4	
-3/2		1		5
-2			1	
-5/2				1

It is apparent from Table 4.1. that for more than two nuclei there will be sub-determinants which cannot be solved explicitly to give the relevant energy levels. In many cases these sub-determinants may be further factorised using either (2) or (3) or both. There are many cases where the use of total spin quantum number factorisation alone, gives the quicker solution to the problems 4 (ii) to (v) although there is usually an equivalent, if laborious, solution using symmetry factorisation [1]. There are some cases where a combination of total spin and symmetry lead to the most efficient solution to the problem [1] and many cases where symmetry alone will give the desired factorisation [14, 15]. The interdependence of these two methods necessitates a detailed consideration of both if the advantages and limitations of the use of symmetry in NMR problems are to be appreciated.

4.2. Total Spin Quantum Number Factorisation

The foundation of the factorisation of the determinant of the Hamiltonian matrix using total spin rests on the fact that there is no mixing between wave functions having a different eigenvalue of I^2 (the square of the total spin angular momentum of the group). The eigenvalues permitted are of the form $\hbar^2 I_T(I_T + 1)$, where I_T is the total spin quantum number associated with I^2. It also follows that there are no transitions allowed between the different eigenstates with different I_T. (Selection rule $\Delta I_T = 0$ applies and I_T is a 'good' quantum number) [26].

The group of nuclei for which this applies behaves like a single *composite particle* which, may exist in different 'spin states' characterised by values of I_T from 0 or I to NI, where I is the spin quantum number. There are $(N + 1)/2$ spin states for N odd and $((N/2) + 1)$ spin states for N even. The magnetic quantum number m may take values I_T, $I_T - 1 \ldots -I_T$ for each value of I_T, giving a total of $(2I_T + 1)$ possible values called the multiplicity of the spin state. The allowed values of I_T for four nuclei are 2, 1 and 0 with multiplicities 5, 3, 1 and labelled Qt, T, S respectively. Each spin state has an associated statistical weight, g, which is the degeneracy of the spin representation in the appropriate symmetry group. The statistical weight is given by

$$g = \frac{(2I_T + 1)N!}{\left(\dfrac{N}{2} - I_T\right)! \left(\dfrac{N}{2} + I_T + 1\right)!} \qquad\qquad 4(i)$$

when the spin state of N nuclei is characterised by the total spin quantum number I_T.

Total spin factorisation may be applied in a limited number of cases only, where the groups of nuclei are magnetically equivalent. These are listed here with the conditions for magnetic equivalence.

4.2.1. Isolated Groups of Nuclei. (Spin I = 1/2)

a) Isotropic media (random molecular motion averages direct coupling).

The nuclei must be isochronous (identical Larmor frequencies). The nuclei are said to be magnetically equivalent even when there is more than one unique coupling constant between them; example – ethylene – one line spectrum. The nuclei are said to be magnetically fully equivalent when there is only one unique coupling constant between them; example – methane – one line spectrum.

b) Anisotropic media (the molecules are partially oriented so that the direct coupling does not average to zero).

The nuclei must have spin 1/2 and be isochronous. There must be only one unique direct coupling constant between the nuclei. It usually follows that there will be only one indirect coupling constant and the nuclei are magnetically fully equivalent [7]. However the direct coupling contributes a first order term to the Hamiltonian matrix elements and N + 1 lines are observed in the spectrum with binomial intensity distribution; example – methane – five lines, 1:4:6:4:1 (if it is possible to orientate methane). Nuclei with spin > 1/2 cannot be magnetically equivalent when included in oriented molecules in anisotropic media [8, 10].

4.2.2. Interacting Groups of Nuclei

The additional terms in the Hamiltonian involve interactions between the groups and for magnetic equivalence of the nuclei of the interacting groups the requirements are as follows.

a) Isotropic media.

The nuclei must be isochronous.

The nuclei of a group of magnetically equivalent nuclei must be equally coupled to each and every nucleus outside the group; example − propane $CH_3 \cdot CH_2 \cdot CH_3$ where internal rotation averages the vicinal coupling constants to a single value − A_6B_2 system.

b) Anisotropic media.

The nuclei must have spin 1/2 and be isochronous.

There must be only one unique direct coupling between the nuclei of the group and only one direct coupling between the nuclei of the group and every other nucleus in the system; example − ethyl iodide $CH_3 \cdot CH_2I$ A_3B_2. The nuclei must be magnetically fully equivalent.

4.2.3. Total Spin and Symmetry

There are some cases where total spin although relevant and valid does not complete the factorisation of the Hamiltonian matrix. These involve mixed cases of magnetic equivalence and non-equivalence, notably the $A_NA'_NBB'$ systems [11]. The two groups A_N and A'_N although internally magnetically equivalent are not themselves magnetically equivalent because there exist two distinguishable coupling constants J_{AB} and $J_{AB'}$. The problem can be constructed simply as the product of the allowed spin states. Specifically for $N = 3$ the product wavefunctions, using the multiplicity notation are

$$(Q + D)_A D_B D_B'(Q + D)_{A'}$$

$Q_A D_B D_B' Q_{A'}$

Q	D	D	D	Q_A is the quartet state of the A group of nuclei $I_T = 3/2$
D	D	D	Q	D is a doublet state $I_T = 1/2$
D	D	D	D	

The wavefunctions are characterised completely here by the appropriate I_T and m but the maximum factorisation is achieved by choosing to take appropriate linear combinations of Q D D' Q' product functions which may be classified as symmetric or antisymmetric with respect to the plane of symmetry (or C_2 axis) which the system as a whole possesses. The examples given in Table 4.2. are for $D_B = D_{B'} = (1/2, 1/2)$ and the (I_T, m) values quoted then refer to the $Q_A Q_{A'}$ states.

Table 4.2. *Symmetric and Antisymmetric* $Q_A Q_A$ *combinations for the* A_3A_3BB' *system*

Symmetric	Antisymmetric
(3/2, 3/2)(3/2, 3/2)	
(3/2, 3/2)(3/2, 1/2) + (3/2, 1/2)(3/2, 3/2)	(3/2, 3/2)(3/2, 1/2)−(3/2, 1/2)(3/2, 3/2)
(3/2, 3/2)(3/2, −1/2) + (3/2, −1/2)(3/2, 1/2)	((3/2, 3/2)(3/2, −1/2)−(3/2, −1/2)(3/2, 1/2))
(3/2, 1/2)(3/2, 1/2)	
(3/2, 1/2)(3/2, −1/2) + (3/2, −1/2)(3/2, 1/2)	((3/2, 1/2)(3/2, −1/2)−(3/2, −1/2)(3/2, 1/2))
(3/2, 3/2)(3/2, −3/2) + (3/2, −3/2)(3/2, 3/2)	((3/2, 3/2)(3/2, −3/2)−(3/2, −3/2)(3/2, 3/2))

The advantages of the composite particle or total spin method include facile construction of wavefunctions and then relatively easy determination of the Hamiltonian matrix leading to rapid completion of the steps outlined at the beginning of this section. However it cannot be employed in systems composed of magnetically non-equivalent nuclei, and it is in these systems where symmetry factorisation can be very useful.

4.3. Symmetry Factorisation of the Determinant of the Hamiltonian Matrix

The factorisation of the determinant of the Hamiltonian matrix using the symmetry properties of the system involves a number of clearly defined steps. These are dealt with in some detail in appendix B and summarised here. Their application is emphasised in examples in this section immediately after their statement given below.

1. recognition of the appropriate permutation group as indicated in section 3, (detail in appendix B).

2. construction or identification of the mathematical representation (character table) composed of a number of *symmetry species* or *irreducible representations*, which defines the allowed ways in which the eigenfunctions of the symmetry operations (permutations) transform under the permutations of the group. An example of the construction of a character table is given in appendix B. It is not necessary to construct the character table in those cases where the system may be considered rigid and where the permutation group of a non-rigid system is recognised as being isomorphic with a rigid group.

3. identification of equivalent sets of basic product wavefunctions which are transformed into one another by the operations of the group for each set of symmetrically equivalent nuclei. The equivalent sets form the basis for a mathematical representation of the system which may be reduced to the sum of irreducible representations using a simple formula which will be introduced later in example [3].

4. the basic symmetry wavefunctions for each group can then be constructed as linear combinations of the basic product wavefunctions which transform properly under the operations of the group. The proper transformation behaviour is assured by the use of a standard group theory formula in the construction of symmetrised wavefunctions [3].

5. construction of the symmetrised wavefunctions for the system as a whole by taking products of the symmetrised wavefunctions and if necessary linear combinations of such products of the group symmetrised wavefunctions, classifying them according to their new transformation properties.

The steps (i) to (v) may appear overwhelming to someone unfamiliar with the language of group theory. The statements need to be couched in the formalism of group theory but are not claimed to be mathematically rigid. The only way to appreciate the implications of these generalisations is to examine examples of their application.

The AA′BB′ System-o-Dichlorobenzene

The permutation group depends upon relevant interactions between the nuclei and the equivalence of nuclei. The molecule may be considered as rigid and the symmetry elements of the configuration averaged over rapid vibrational motions [29] include the plane of the benzene ring, σ, the plane of symmetry perpendicular to the benzene ring, σ', and a two fold axis C_2 defined by the intersection of the two planes.

The interest in this molecule from the point of view of NMR lies in the proton spectrum only since the chlorine nuclei are effectively inactive. The NMR parameters v_1, v_2, J_{12}, J_{34}, $J_{13,24}$ and $J'_{14,23}$ are not permuted by the plane of symmetry. The effect of σ' is wholly equivalent, in this context, to C_2 and therefore the effective group can be recognised as either C_s or C_2 which are isomorphic.

The permutations of the group are E (identity) and (12)(34). The NMR system is composed of two groups of nuclei comprising two isochronous nuclei. each. The method of approach used here is to construct symmetrised wavefunctions for each group and then produce symmetrised wavefunctions for the whole system by multiplying the wavefunctions of the two groups together. [2, 12, 13].

Two Equivalent Nuclei with C_s Effective Symmetry

The basic product wavefunctions for two nuclei (spin 1/2) are $\alpha\alpha$, $\alpha\beta$, $\beta\alpha$, $\beta\beta$. The permutation properties of these basic product wavefunctions under the operations of the group are illustrated in Table 4.3. The character table for the C_s point group is given in Table 4.4.

Table 4.3	
E	(12)
$\alpha\alpha$	$\alpha\alpha$
$\alpha\beta$	$\beta\alpha$
$\beta\alpha$	$\alpha\beta$
$\beta\beta$	$\beta\beta$

Table 4.4		
C_s	E	(12)
A	1	1
B	1	-1

It can be seen from the permutation properties of the basic product wavefunctions that $\alpha\alpha$ and $\beta\beta$ are not affected by (invariant under) the permutations of the group (in fact both form a basis for, and belong to, the totally symmetric representation A). However $\alpha\beta$ and $\beta\alpha$ form a two dimensional equivalent set from which it is possible to construct two symmetrised wavefunctions. It is necessary to establish which symmetry species these symmetrised wavefunctions may belong to and this can be achieved directly from a group theoretical formula.

$$n^\gamma = \frac{1}{h} \sum_c g_c \chi_c \chi_c^{(\gamma)} \qquad \qquad …4\,(ii)$$

This formula gives the number of symmetrised wavefunctions (on degenerate sets of wavefunctions) in each symmetry species γ. The total number of symmetry operations in the group is h, g_c is the number of symmetry operations for each *class* C of the group, χ_c is the number of spin product wavefunctions which are not changed (invariant) under the symmetry operations in each class C, and $\chi_c^{(\gamma)}$ are the standard symmetry species characters which are tabulated (Appendix C) or can be derived for each group.

In this case the total number of symmetry operations, is two and the number of spin product wavefunctions which are not changed are given in Table 4.5.

Table 4.5

	E	(12)
g_c	1	1
χ_c	2	0

The number of symmetrised wavefunctions which can be constructed from $\alpha\beta$ and $\beta\alpha$ belonging to the symmetry species A is given by

$$n(A) = 1/2\,(1 \times 2 \times 1 + 1 \times 0 \times 1) = 1$$

and

$$n(B) = 1/2\,(1 \times 2 \times 1 + 1 \times 0 \times -1) = 1\,.$$

The representation for the equivalent set of $\alpha\beta$ and $\beta\alpha$, $\Gamma = A + B$.

There will be a total of three A species symmetrised wavefunctions (including $\alpha\alpha$ and $\beta\beta$) and one belonging to the B symmetry species.

Note here that total spin gives the same result with different labels. The allowed values of I_T are 1 and 0, the multiplicities of these states are three and one respectively. The labels are I_T and m for total spin factorisation and A, B and m for symmetry factorisation. However total spin is not appropriate for the AA'BB' problem because the nuclei of the groups are magnetically non-equivalent. The symmetrised wavefunctions, which are relevant, may be constructed using, once again, a standard group theoretical formula

$$\psi^\gamma = \eta \sum_p \chi_p^\gamma P\varphi_1 \qquad\qquad \dots 4(\text{iii})$$

ψ^γ is the symmetrised wavefunction which is a linear combination of spin product wavefunctions generated from one of them ($\alpha\beta$ here for example) by the operation of the permutations P (symmetry operation) of the group on it. χ_p^γ is the group character for the operation P appropriate for the symmetry species γ and η is a normalising factor.

The A species symmetrised wavefunction can be constructed from $\alpha\beta$ as follows

$$\psi^A = \eta\,(1 \times E(\alpha\beta) + 1 \times (12)(\alpha\beta)) = \eta\,(\alpha\beta + \beta\alpha)\,.$$

The value of η is given by the requirement that the sum of the squares of the coefficients of the spin product wavefunctions in the symmetrised wavefunction be equal to one [3].

Therefore $2\eta^2 = 1$ and $\eta = \dfrac{1}{\sqrt{2}}$

$$\psi^A = \dfrac{1}{\sqrt{2}}\,(\alpha\beta + \beta\alpha)$$

Similarly

$$\psi^B = \dfrac{1}{\sqrt{2}}\,(\alpha\beta - \beta\alpha)$$

The symmetrised wavefunctions for two nuclei are therefore

A	B
$\alpha\alpha$	
$(\alpha\beta + \beta\alpha)/\sqrt{2}$	$(\alpha\beta - \beta\alpha)/\sqrt{2}$
$\beta\beta$	

schematically represented by

A	B
1	
1	1
1	

Symmetrised Wavefunctions for the AA'BB' System

The symmetrised wavefunctions for the AA'BB' system can be obtained as direct products of the symmetrised wavefunctions for the A_2 and B_2 groups of nuclei. The symmetry of the product wavefunctions can be established by multiplying the group characters for each class of operations as follows

	E	(12)	
$\psi^A(A_2)$	1	1	$A \times A$ product
$\psi^A(B_2)$	1	1	
$\psi^A(AA'BB')$	1	1	A species

	E	(12)	
$\psi^A(A_2)$	1	1	$A \times B$ product
$\psi^B(B_2)$	1	-1	
$\psi^B(AA'BB')$	1	-1	B species

Summarising, the symmetry of the product wavefunctions for the AA'BB' system are given by

	A	B
A	A	B
B	B	A

9*

and these can be used in constructing the energy level diagram as follows:—

Table 4.6. *Multiplication of A_2 and B_2 wavefunctions to give the wavefunctions for the $AA'BB'$ system*

ψA		ψB		$\psi AA'BB'$			
A	B	A	B	AA	BB	AB	BA
				1			
1		1		11		1	1
1	1 ×	1	1 →	111	1	1	1
1		1		11		1	1
				1			

AA and BB products have the same symmetry A, and the AB, BA products have symmetry B. There will be off-diagonal elements between symmetrised wavefunctions of the same symmetry and the same total magnetic quantum number m_T leading to sub-matrices of order 2 and 4 as shown below.

Table 4.7. *The schematic energy level diagram for the $AA'BB'$ system*

m_T	A species	B species
2	1	
1	2	2
0	4	2
-1	2	2
-2	1	

The problem is further simplified in the approximation that the chemical shift between the two types of nuclei is very much greater than the coupling constants between the groups, the $AA'XX'$ system, when off-diagonal elements between wavefunctions with different m_A and m_X may be neglected. This case is considered in more detail in section 5.

The $AA'BB'$ case considered above is one of the simplest systems which can be used to illustrate the application of group theory in NMR. It has served to introduce the ideas and theorems at a simple level but in many ways it is only superficial in illustrating the problems which may be encountered. Specifically, the case where the symmetrised wavefunctions of a group include wavefunctions which belong to a degenerate symmetry species as in $AA'A''BB'B''$ and $AA'A''A'''BB'B''B'''$ systems has not been illustrated.

The AA′A″BB′B″ System

A model for this system is symtrifluorobenzene

$$J_{15} = J_{16} = J_{24} = J_{26} = J_{34} = J_{35}$$
$$J_{14} = J_{25} = J_{36}$$

The notation AA′A″BB′B″ implies that there are two groups of three magnetically non-equivalent nuclei and has been adopted for symmetry reasons only. This notation would only have significance in NMR if the two groups of nuclei were strongly coupled (see page 112). The appropriate NMR notation at commercially available magnetic field values is AA′A″XX′X″ and the implications of this notation are considered in section 5.

The Permutation Group

The permutation group consists of six permutations

E, (123)(456)   (12)(45)
   (132)(465),   (13)(46)
                   (23)(56),

which can be identified with the symmetry elements of the D_3 symmetry group

E, C_3 (clockwise) C_2 (3,6)
                           C_2 (2,5)
   C_3 (anticlockwise), C_2 (1,4),

and indeed the NMR permutation group is isomorphic with D_3. The character table is given below

Table 4.8. *Character table for the D_3 symmetry group*

D_3	E	$2C_3$	$3C_2$
A_1	1	1	1
A_2	1	1	−1
E	2	−1	0

The symmetry wavefunctions for the system can be constructed from products of the group symmetry wavefunctions $\psi(A_3)$ and $\psi(B_3)$. The group symmetry wavefunctions $\psi(A_3)$ and $\psi(B_3)$ can be classified in equivalent ways since it can be seen from the permutation operations of the group that the numbering system has been chosen in a cyclic manner such that the A nuclei and B nuclei are transformed in equivalent ways. Therefore it is necessary to consider one group only.

Symmetrised Wavefunctions for Three Equivalent Nuclei

The $\alpha\alpha\alpha$ spin product wavefunction can be seen to constitute a set, as does $\beta\beta\beta$, and both transform according to the A_1 symmetry species.

	E	$2C_3$	$3C_2$
$\alpha\alpha\alpha$	$\alpha\alpha\alpha$	$\alpha\alpha\alpha$	$\alpha\alpha\alpha$

The spin product wavefunctions with $m_T = +1/2$ $\alpha\alpha\beta$, $\alpha\beta\alpha$, $\beta\alpha\alpha$ form an equivalent set with transformation properties equivalent to the equivalent set with $m_T = -1/2$ $\beta\beta\alpha$, $\beta\alpha\beta$, $\alpha\beta\beta$. Therefore to construct the symmetrised wavefunctions for three spin 1/2 nuclei it is sufficient simply to consider the transformation properties of one equivalent set in detail. There are shown below in Table 4.9.

Table 4.9. *Transformation Properties of the* $\beta\alpha\alpha \equiv [1]$, $\alpha\beta\alpha \equiv [2]$, *and* $\alpha\alpha\beta \equiv [3]$ *spin product wavefunctions for three nuclei under the operations of the* D_3 *symmetry group*

	E	(123)	(132)	(12)	(13)	(23)
	[1]	[2]	[3]	[2]	[3]	[1]
	[2]	[3]	[1]	[1]	[2]	[3]
	[3]	[1]	[2]	[3]	[1]	[2]
χ_c	3	0	0	1	1	1

The numbers are used to locate the position of the β spin function in the product wavefunction, $\beta\alpha\alpha \equiv [1]$.

The number of wavefunctions which can be constructed belonging to the symmetry species of the group can be calculated using eq. 4 (ii).

$$n^{A_1} = 1/6 \ (1 \times 3 \times 1 + 2 \times 0 \times 1 + 3 \times 1 \times 1) = 1$$
$$n^{A_2} = 1/6 \ (1 \times 3 \times 1 + 2 \times 0 \times 1 - 3 \times 1 \times 1) = 0$$
$$n^{E} \ = 1/6 \ (1 \times 3 \times 2 - 2 \times 0 \times 1 + 3 \times 1 \times 0) = 1$$

The representation corresponding to the transformation properties of the three product wavefunctions can therefore be reduced to the sum of A_1 and E only.

or $\Gamma(m_T = \pm 1/2) = A_1 + E$ (three symmetrised wavefunctions)

The wavefunctions which belong to the E classification are a doubly degenerate pair. They correspond to the same energy but are orthogonal i.e. the sum of the products of the coefficients of the spin product wavefunctions in the symmetrised wavefunctions is zero – thus

$$\psi_a^E = C_{1a}\alpha\alpha\beta + C_{2a}\alpha\beta\alpha + C_{3a}\beta\alpha\alpha$$
$$\psi_b^E = C_{1b}\alpha\alpha\beta + C_{2b}\alpha\beta\alpha + C_{3b}\beta\alpha\alpha$$
$$C_{1a}C_{1b} + C_{2a}C_{2b} + C_{3a}C_{3b} = 0, \text{ for } \psi_a^E$$

and ψ_b^E to be orthogonal.

The symmetrised wavefunctions can be constructed using eq. 4 (iii) and, in the case of ψ_b^E, the orthogonality condition

$$\psi^{A_1} = \eta \sum_p \chi_p^{A_1} P[1] = \frac{1}{\sqrt{3}}([1] + [2] + [3])$$

$$\psi_a^E = \eta \sum_p \chi_p^E P[1] = \frac{1}{\sqrt{6}}(2[1] - [2] - [3])$$

ψ_b^E can be constructed from a linear combination of the functions generated by using [1] and [2] in eq. 4 (iii) chosen such that ψ_b^E is orthogonal to ψ_a^E, leaving normalisation until last (denoted by $\psi_{a'}^E$ primed notation)

$$X \psi_{a'}^E[1] + Y \psi_{a'}^E[2] = X[2[1] - [2] - [3]] + Y[2[2] - [1] - [3]]$$
$$= (2X - Y)[1] + (2Y - X)[2] - (X + Y)[3]$$
$$\psi_{a'}^E = 2[1] - [2] - [3]$$

Hence $2(2X - Y) - (2Y - X) + (X + Y) = 0$ (orthogonality condition)

$$6X - 3Y = 0$$
$$X = \frac{Y}{2}$$
$$\psi_{b'}^E = X(2[1] - [2] - [3]) + 2X(2[2] - [1] - [2])$$
$$= 3X([2] - [3]) \qquad X = \tfrac{1}{3}\sqrt{2}$$
$$\psi_b^E = \frac{1}{\sqrt{2}}([2] - [3]) \qquad \text{normalised}$$

The symmetrised wavefunctions for three spin 1/2 nuclei can be summarised as: –

A_1 species	E species
[0]	
$([1] + [2] + [3])(\sqrt{3})^{-1}$	$(2[1] - [2] - [3])(\sqrt{6})^{-1}, ([2] - [3])(\sqrt{2})^{-1}$
$([12] + [13] + [23])(\sqrt{3})^{-1}$	$(2[12] - [13] - [23])(\sqrt{6})^{-1},$
[123]	$([13] - [23])(\sqrt{2})^{-1}$

or, alternatively, the schematic energy level diagram can be written:

	A_1 species	E species
	1	
	1	1
	1	1
	1	
I_T	3/2	1/2

The total spin classification is also given here to show the one to one correspondence between the symmetry and I_T factorisation, for this isolated group [30].

Symmetrised Wavefunctions for the AA'A''BB'B'' System

The symmetrised wavefunctions for the AA'A''BB'B'' system can be constructed from products of the symmetrised wavefunctions of the two groups.

The symmetrised wavefunctions for the A group of nuclei have already been found and the B nuclei symmetrised wavefunctions can be written down by analogy, except that one has to be careful to choose the signs so that they transform in exactly the same way as the A nuclei wavefunctions [29].

The products can be classified as

$$A_1 \times A_1, \; A_1 \times E, \; E \times A_1 \quad \text{and} \quad E \times E.$$

The $A_1 \times A_1$, $A_1 \times E$, and $E \times A_1$ products can be shown by simple multiplication of the characters for each class of operation to have A_1, E and E symmetry respectively. For example

	E	$2C_3$	$3C_2$	
A_1	1	1	1	
E	2	-1	0	
$A_1 \times E$	2	-1	0	(transforms as E)

The $E \times E$ products however present a new problem since the product of the respective characters gives: —

	E	$2C_3$	$3C_2$
$E \times E$ products	4	1	0

The representation given immediately above which is a mathematical formalism of the transformation properties of the $E \times E$ products under the operations of the D_3 group can be seen not to occur as such in the D_3 group character table. However, it can be shown that this representation is simply the sum of irreducible representations of the D_3 symmetry group.

$$E \times E = A_1 + A_2 + E$$

This means that from the products of the four E species wavefunctions $\psi_a^E(A)$, $\psi_a^E(B)$, $\psi_b^E(A)$ and $\psi_b^E(B)$ of the type $\psi^E(A)$. $\psi^E(B)$ four wavefunctions can be constructed which have the symmetry A_1, A_2 and E (two of a degenerate pair).

The $E \times E$ representation can be reduced to $A_1 + A_2 + E$ using the formula

$$N^\gamma = \frac{1}{h}\sum_c g_c \, \chi(\text{product})\chi_c^\gamma \qquad\qquad 4\,(iv)$$

where N^γ is the number of wavefunctions of species γ which can be constructed from products of the type considered, h and g_c have the same meaning as in eq. 4(ii)χ(product) as the characters derived by multiplying the species characters of the product functions together and the χ^γ are the characters of the symmetry species γ.

$$N^{A_1} = 1/6 \, (1 \times 4 \times 1 + 2 \times 1 \times 1 + 3 \times 0 \times 1) = 1$$
$$N^{A_2} = 1/6 \, (1 \times 4 \times 1 + 2 \times 1 \times 1 + 3 \times 0 \times (-1)) = 1$$
$$N^{E} = 1/6 \, (1 \times 4 \times 1 + 2 \times 1 \times (-1) + 3 \times 0 \times 0) = 1$$

The appropriate linear combinations of the four $E \times E$ products of ψ_a^E (A), ψ_b^E (A), ψ_a^E (B) and ψ_b^E (B) follow from the application of eq. 4(iii) to

ψ_a^E (A) ψ_a^E (B) when two wavefunctions result

$$\frac{1}{\sqrt{2}} [\psi_a^E (A) \, \psi_a^E (B) + \psi_b^E (A) \, \psi_b^E (B)] = \psi^{A_1} AA'A''BB'B''$$

$$\frac{1}{\sqrt{2}} [\psi_a^E (A) \, \psi_a^E (B) - \psi_b^E (A) \, \psi_b^E (B)] = \psi_a^E AA'A''BB'B''$$

and to ψ_a^E (A) ψ_b^E (B) when two more

$$\frac{1}{\sqrt{2}} [(\psi_a^E (A) \, \psi_b^E (B) - \psi_a^E (A) \, \psi_b^E (B)] = \psi^{A_2} AA'A''BB'B''$$

$$\frac{1}{\sqrt{2}} [\psi_a^E (A) \, \psi_b^E (B) + \psi_b^E (A) \, \psi_a^E (B)] = \psi_b^E AA'A''BB'B''$$

are the result [29].

The factorisation of the secular determinant depicted in Table 4.10 is a simplification of the initial binomial factorisation based on m_T. Further factorisation can be achieved using the mathematical approximation already mentioned. This is given in section 5.

AA'A"A'''BB'B"B''' Systems

The diversity of permutation groups to which these systems can belong has been mentioned in section 3.3.2. Each of the AA'A"A'''BB'B"B''' systems with T_d, D_4, D_{2h}, D_{2d} and C_{2h} symmetry has been studied using the methods described above [15]. One of the interesting points which arises is that the A_4 and B_4 nuclei incorporated in the complete systems do not transform in equivalent ways under the operations of the D_{2h}, D_{2d} and D_4 symmetry groups. No new principle arises however in studying these systems and therefore the results are presented here as further examples of the simplification which can be achieved by the use of symmetry.

AA'A"A'''XX'X"X''' Systems with T_d Symmetry [31, 32, 33]

The classifications of the wavefunctions of the two sets of nuclei are identical

$$\Gamma = 5A_1 + E + 3T_2 \quad \text{(T implies three fold degeneracy)}$$

Table 4.10. *Derivation of the schematic factorisation of the determinant for the* $AA'A''BB'B''$
system with D_3 *symmetry*

M_T	A_1	E	A_1	E	A_1A_1	$EE(A_1)$	$EE(A_2)$	AE	EA	$(EE)E$
3					1	1				
5/2										
2					1 1	2		1	1	2
3/2	1		1							
1					1 1 1	1 4	1 1	1 1	1 1	1 5
1/2	1	1	1	1						
0					1 1 1 1	1 1 6	1 1 2	1 1	1 1	1 1 6
−1/2	1	1	1	1						
−1					1 1 1	1 4	1 1	1 1	1 1	1 5
−3/2	1		1							
−2					1 1	2		1	1	2
−5/2										
−3					1	1				

This can be compared directly with the total spin classification –

$$\Gamma = 1\,\mathrm{Qt}\,(I_T = 2) + 3\,\mathrm{T}\,(I_T = 1) + 2\,\mathrm{S}\,(I_T = 0).$$

and it can be seen that symmetry and total spin factorisation are identical. However, it is necessary to classify the wavefunctions and construct them according to their symmetry properties in order to set up the $AA'A''A'''BB'B''B'''$ T_d system Hamiltonian matrix.

The cross-products of the wavefunctions of the two sets lead to product wavefunctions of the $AA'A''A'''BB'B''B'''$ system with symmetry indicated in Table 4.11.

Table 4.11. *Symmetry of the product wavefunctions* $\psi(A_4) \times \psi(B_4)$ *in the* T_d *symmetry group*

	$A_1\,(5)$	$E\,(1)$	$T_2\,(3)$
$A_1\,(5)$	$A_1\,(25)$	$E\,(5)$	$T_2\,(15)$
$E\,(1)$	$E\,(5)$	$(A_1 + A_2 + E)\,(1)$	$(T_1 + T_2)\,(3)$
$T_2\,(3)$	$T_2\,(15)$	$(T_1 + T_2)\,(3)$	$(A_1 + E + T_1 + T_2)\,(9)$

The symmetry wavefunctions of the $AA'A''A'''BB'B''B'''$ (T_d) system can be summarised as

$$\Gamma_{T_d} = 35\,A_1 + A_2 + 20\,E + 15\,T_1 + 45\,T_2$$

This factorisation is illustrated schematically in Table 4.12.

Table 4.12. *Schematic illustration of the symmetry factorisation for the* $AA'A''A'''BB'B''B'''$ *system with* T_d *symmetry*

	A_1	A_2	E	T_1	T_2
M_T					
4	1				
3	2				2
2	4		3	1	5
1	6		4	4	10
0	9	1	6	5	11
-1	6		4	4	10
-2	4		3	1	5
-3	2				2
-4	1				

The symmetry factorisation for the other $AA'A''A'''BB'B''B'''$ systems is summarised in each case below as the linear sum of the wavefunctions with appropriate symmetry.

Summary of the Factorisation Achieved Using Symmetry in $AA'A''A'''BB'B''B'''$ Systems

$$D_4 \begin{cases} \Gamma(A_4) & = 6A_1 + B_1 + 3B_2 + 3E \\ \Gamma(B_4) & = 6A_1 + 3B_1 + B_2 + 3E \\ \Gamma(AA'A''A'''BB'B''B''') & = 51A_1 + 19A_2 + 33B_1 + 33B_2 + 60E \end{cases}$$

$$D_{2h} \begin{cases} \Gamma(A_4) & = 7A_g + 3B_{3g} + 3B_{1u} + 3B_{2u} \\ \Gamma(B_4) & = 7A_g + 3B_{2g} + 3B_{1u} + 3B_{3u} \\ \Gamma(AA'A''A'''BB'B''B''') & = 58A_g + 18B_{1g} + 30B_{2g} + 30B_{3g} \\ & \quad + 18A_u + 42B_{1u} + 30B_{2u} + 30B_{3u} \end{cases}$$

$$D_{2d} \begin{cases} \Gamma(A_4) & = \\ \Gamma(B_4) & = \end{cases} \Big\} 6A_1 + B_1 + 3B_2 + 3E$$
$$\qquad\ \Gamma(AA'A''A'''BB'B''B''') = 55A_1 + 15A_2 + 1B_1 + 45B_2 + 60E$$

$$C_{2h} \begin{cases} \Gamma(A_4) & = \\ \Gamma(B_4) & = \end{cases} \Big\} 7A_g + 3B_g + 3A_u + 3B_u$$
$$\qquad\ \Gamma(AA'A''A'''BB'B''B''') = 76A_g + 60A_u + 60B_g + 60B_u$$

5. The X Approximation

The examples which have been given thus far have emphasised the role of symmetry in simplifying the problems encountered in the analysis of NMR spectra. However further simplification can be achieved in those cases where off-diagonal elements of the Hamiltonian matrix are small compared with the difference between the corresponding diagonal elements. In general this implies that coupling constants between groups of nuclei are small compared with the chemical shift difference between the nuclei of the respective groups. The groups of nuclei are then said to be weakly coupled and this is reflected in the NMR notation by denoting the groups by letters well-separated in the alphabet (12). The AA'BB' system may be denoted AA'XX' if the chemical shift can be increased (by using a high magnetic field − superconducting magnets, for example) such that J_{AB}, $J_{AB'} \ll |v_A - v_B|$. In these cases the Hamiltonian matrix can be derived in diagonal terms only using second-order perturbation theory and where the second-order perturbation contributions of the type $J_{AB}^2/(v_A - v_B)^2$ are negligible the term X approximation is used [2, 12]. The conditions for the validity of the X approximation must of course be modified depending upon the Hamiltonian which is used. When direct coupling between nuclei is included the X approximation, although valid for the scalar coupling term, need not be valid so that in these cases the additional simplification may not be achieved.

5.1. Homonuclear Systems of Nuclei

a) Systems Composed of Protons Only

The X approximation may or may not be valid in systems composed of protons only depending upon the relative magnitudes of the coupling constants and chemical shifts [2, 12]. The proton spectrum of pyridine in the isotropic phase at 14,000 gauss can be described to a good approximation by the notation ABB'XX' where the protons ortho to the nitrogen atom, XX', are appreciably less shielded than the meta and para protons, ABB' [19, 20]. However, when pyridine is partially oriented as solute in a liquid crystal the notation must be amended to ABB'CC' because the direct coupling averaged over the restricted orientations of the molecules with respect to the magnetic field are very much larger than the scalar couplings and cannot be neglected in off-diagonal matrix elements [21]. The appropriate factorisation of the secular determinant in these two cases are compared in Table 5.1. The total magnetic quantum numbers of the ortho protons m(XX') and the meta-para protons m(ABB') provide the basis for the additional factorisation where the X approximation is valid.

b) Systems of Nuclei Other than Protons

The chemical shifts in system composed of nuclei other than protons are normally large and the X approximation is valid in these cases. Most penta-fluorophenyl derivatives yield spectra which can be analysed as AA'MM'X systems with well separated bands due to the fluorine nuclei ortho meta and para to the substituent [22]. However there are exceptions and pentafluoroanisole is best described as an ABB'XX' system at 14,000 gauss in isotropic media [23]. The classification of these systems in liquid crystal solvents cannot be easily generalised.

Table 5.1. *A comparison of the factorisation of the determinant of the Hamiltonian matrix for pyridine in a) isotropic phase, b) liquid crystal*

a) Isotropic phase ABB'XX'						b) Liquid crystal phase ABB'CC'	
C_2	A			B		A	B
m_T							
5/2	1					1	
3/2	2	1		1	1	3	2
1/2	2	3	1	1	3	6	4
−1/2	1	3	2	3	1	6	4
−3/2		1	2	1	1	3	2
−5/2			1			1	
m(XX')	1	0	−1	1	0	−1	

5.2. Heteronuclear Systems of Nuclei

The chemical shifts between nuclei of different isotopic species are of the order of megacycles at commercially available values of the magnetic field. The X approximation is valid in such systems both for the isotropic phase and liquid crystal spectra. Examples of these systems include 1,1-difluoroethylene [13], symmetrical trifluorobenzene [14], 1,4 difluorobenzene and 2,3,5,6-tetrafluorobenzene [24]. The completed factorisation of the secular determinants in these cases is given in Tables 5.2, 5.3 and 5.4 respectively. Table 5.5 summarises the factorisation which can be achieved for the tetrafluorocyclobutane AA'A"A'''XX'X"X''' (see footnote to Table 5.5) -type systems, using symmetry and the X approximation.

Table 5.2. *Factorisation of the determinant for the AA'XX' system with C_2 symmetry*

C_2	A			B		
m_T						
2	1					
1	1	1		1	1	
0	1	2	1		2	
−1		1	1		1	1
−2			1			
m(XX')	1	0	−1	1	0	−1

Table 5.3. *Factorisation of the determinant for the $AA'A''XX'X''$ system with D_3 symmetry*

D_3	A_1				A_2		E			
m_T										
3	1									
2	1	1					1	1		
1	1	2	1		1		1	3	1	
0	1	2	2	1	1	1		3	3	
−1		1	2	1		1		1	3	1
−2			1	1					1	1
−3				1						
$m(XX'X'')$	3/2	1/2	−1/2	−1/2	1/2	−1/2	1/2	1/2	−1/2	−3/2

Table 5.4. *Factorisation of the determinant for the $AA'XX'X''X'''$ system with D_2 symmetry*

D_2	A_1			B_1			B_2			B_3		
m_T												
3	1											
2	1	1		1			1	1		1		
1	3	2	1	1	2		1	2		1	2	
0	1	4	1	1	2	1	1	4	1	1	2	1
−1	1	2	3		2	1		2	1		2	1
−2		1	1		1			1	1			1
−3			1									
$m(AA')$	1	0	−1	1	0	−1	1	0	−1	1	0	−1

Table 5.5. *Basic molecular symmetry wavefunction diagram*
(i) D_{2h} symmetry

		a) Cyclobutane derivative					b) 1:4 Cyclohexadiene derivative				
	$m(A)_4$	2	1	0	−1	−2	2	1	0	−1	−2
Species	$m[AX]_4$										
	4	1					1				
	3	1	1				1	1			
	2	3	2	3			3	2	3		
	1	1	4	4	1		1	4	4	1	
Ag	0	1	2	10	2	1	1	2	10	2	1
	−1		1	4	4	1		1	4	4	1
	−2			3	2	3			3	2	3
	−3				1	1				1	1
	−4					1					1

Table 5.5 (continued)

(i) D_{2h} symmetry

Species		a) Cyclobutane derivative					b) 1:4 Cyclohexadiene derivative				
	$m(A)_4$	2	1	0	-1	-2	2	1	0	-1	-2
A_u	2	2					2				
	1	2	2				2	2			
	0	2	2	2			2	2	2		
	-1		2	2				2	2		
	-2			2					2		
Species	$m[AX]_4$										
B_{1g}	3	1									
	2	2	1				2				
	1	4	2	1			2	2			
	0		2	4	2		2	2	2		
	-1			1	2	4		2	2		
	-2				1	2			2		
	-3					1					
B_{1u}	3	1					1	1			
	2	1	2				1	2	1		
	1	1	2	4			1	4	4	1	
	0		2	4	2			2	6	2	
	-1			4	2	1		1	4	4	1
	-2				2	1			1	2	1
	-3					1				1	1
Species	$m[AX]_4$										
B_{2g}	3	1					1				
	2	1	2				2	1			
	1	1	2	4			4	2	1		
	0		2	4	2			2	4	2	
	-1			4	2	1			1	2	4
	-2				2	1				1	2
	-3					1					1
B_{2u}	3	1					1				
	2	2	1				1	2			
	1	4	2	1			1	2	4		
	0		2	4	2			2	4	2	
	-1			1	2	4			4	2	1
	-2				1	2				2	1
	-3					1					1

Table 5.5 (continued)

(ii) D_4 symmetry

Species m(A)$_4$	A_1 2	1	0	−1	−2	A_2 2	1	0	−1	−2	B_1 2	1	0	−1	−2
m[AX]$_4$															
4	1														
3	1	1									1				
2	2	2	2			2					1	2	1		
1	1	4	4	1		2	2				1	2	4		
0	1	2	8	2	1	2	3	2				2	5	2	
−1		1	4	4	1		2	2					4	2	1
−2			2	2	2			2					1	2	1
−3				1	1										1
−4					1										

Species m(A)$_4$	B_2 2	1	0	−1	−2	E 2	1	0	−1	−2
m[AX]$_4$										
3		1				1	1			
2	1	2	1			1	4	1		
1		4	2	1		1	6	6	1	
0		2	5	2			4	8	4	
−1		1	2	4			1	6	6	1
−2			1	2	1			1	4	1
−3				1					1	1

(iii) $D_{2d}(C_{4v})$ symmetry

Species m(A)$_4$	A_1 2	1	0	−1	−2	A_2 2	1	0	−1	−2	B_1 2	1	0	−1	−2
m[AX]$_4$															
4	1														
3	1	1													
2	2	3	2			1					1	1	1		
1	1	4	4	1		2	2					2	2		
0	1	3	7	3	1	1	3	1				1	5	1	
−1		1	4	4	1		2	2					2	2	
−2			2	3	2			1					1	1	1
−3				1	1										
−4					1										

Table 5.5 (continued)

(iii) $D_{2d}(C_{4v})$ symmetry

Species m(A)$_4$	B_2 2	1	0	-1	-2	E 2	1	0	-1	-2
m[AX]$_4$										
3	1	1				1	1			
2	1	3	1			1	4	1		
1	1	4	4	1		1	6	6	1	
0		3	5	3			4	8	4	
-1		1	4	4	1		1	6	6	1
-2			1	3	1			1	4	1
-3				1	1				1	1

(iv) $C_{2h}(C_{2v})$ symmetry

Species m(A)$_4$	A_g 2	1	0	-1	-2	B_g 2	1	0	-1	-2	A_u 2	1	0	-1	-2
m[AX]$_4$															
4	1														
3	1	1				1	1				1	1			
2	3	4	3			1	4	1			1	4	1		
1	1	6	6	1		1	6	6	1		1	6	6	1	
0	1	4	12	4	1		4	8	4			4	8	4	
-1		1	6	6	11		1	6	6	1		1	6	6	1
-2			3	4	3			1	4	1			1	4	1
-3				1	1				1	1				1	1
-4					1										

Species m(A)$_4$	B_u 2	1	0	-1	-2
m[AX]$_4$					
4					
3	1	1			
2	1	4	1		
1	1	6	6	1	
0		4	8	4	
-1		1	6	6	1
-2			1	4	1
-3				1	1
-4					

Another notation for AA′A″A‴XX′X″X‴ systems has been suggested which replaces the primes by square brackets to denote magnetic non-equivalence, thus [AX]$_4$. This has been introduced here because of space problems in the tables but it is not known to be in general use yet [25, 31].

5.3. Sub-spectral Analysis

One of the most important results which has come from the use of the total magnetic quantum, the total spin quantum number, symmetry and the X approximation in factorising the determinant of the Hamiltonian matrix has been the recognition of sub-patterns of energy levels, and the transitions between them, as identifiable *sub-spectra* with all the features characteristic of analogous isolated systems of nuclei [26].

The AA′XX′ System

The $1:1:1$ (A species) and 1 (B species) sub-patterns corresponding to $m(XX') = \pm 1$ in Table 5.2 have been identified as a_2 sub-spectra with effective chemical shifts —

$$\nu_a = \nu_A \pm 1/2\,(J_{AX} + J_{AX'}).$$

The $1:2:1$ (A and B species) sub-patterns have been proved to be characterised by a Hamiltonian analogous to the AB system and the corresponding sub-spectral parameters which describe the ab subspectra are —

A species
$$\nu_a = \nu_A + 1/2\,(J_{AX} - J_{AX'})$$
$$\nu_b = \nu_A - 1/2\,(J_{AX} - J_{AX'})$$
$$J_{ab} = J_{AA'} + J_{XX'}$$

B species
$$\nu_a = \nu_A + 1/2\,(J_{AX} - J_{AX'})$$
$$\nu_b = \nu_A - 1/2\,(J_{AX} - J_{AX'})$$
$$J_{AB} = J_{AA'} - J_{XX'}$$

Summarising, the A and X parts of the AA′XX′ system can be written as the superposition of four sub-spectra

$$AA'\,[AA'XX']_{m(xx')} \equiv [a_2]_{+1} + [a_2]_{-1}$$
$$+ [ab]_0 \qquad A \text{ species}$$
$$+ [ab]_0 \qquad B \text{ species}$$

The AA′A″XX′X″ System

The $[A]_3$ and $[X]_3$ parts of AA′A″XX′X″ systems can be expressed as the superposition of six sub-spectra [14, 27].

$$AA'A''\,[AA'A''XX'X'']_{mxx'x''} \equiv$$
$$[a_3]_{\pm 3/2} \qquad (1:1:1:1,\ A_1 \text{ and } 1:1,\ E)$$
$$[ab_2]_{\pm 1/2} \qquad (1:2:2:1,\ A_1 \text{ and } 1:1,\ A_2)$$
$$[abc]_{\pm 1/2} \qquad (1:3:3:1,\ E)$$

The sub-spectral parameters have been derived in each case [27].

$$[a_3]_{\pm 3/2} \qquad \nu_a = \nu_A \pm 1/2\,(2J_{AX} + J_{AX'})$$
$$[ab_2]_{\pm 1/2} \qquad \nu_a = \nu_A \pm 1/2\,(2J_{AX} - J_{AX'})$$
$$\nu_b = \nu_b \pm 1/2\,J_{AX'}$$
$$J_{ab} = J_{AA'} + J_{XX'}$$
$$[abc]_{\pm} \qquad \nu_a = \nu_A \pm 1/2\,(2J_{AX} - J_{AX'})$$
$$\nu_b = \nu_c = \nu_A \pm 1/2\,J_{AX'}$$

$$J_{ab} = J_{AA'} + 1/2\,(\sqrt{3} - 1)\,J_{XX'}$$
$$J_{ac} = J_{AA'} - 1/2\,(\sqrt{3} + 1)\,J_{XX'}$$
$$J_{bc} = J_{AA'} - 1/2\,J_{XX'}$$

The sub-spectral parameters for the XX'X'' part are identical except in the effective coupling constants within the abc sub-spectra where the appropriate parameters can be generated by writing X for A.

The AA'A''A'''XX'X''X''' Systems

It is possible to pick out a number of sub-patterns of energy levels which bear a superficial resemblance to known systems [15]. However of these only those which can be labelled as a_4 behave like true sub-spectra. It has not been possible to derive meaningful transformations for sub-patterns of greater complexity in such systems.

Appendix A

Matrices and Vectors

Matrix and vector notation are of considerable importance in illustrating the concepts of group theory. The features of matrices and vectors relevant to the problems encountered in group theory are summarised here but no claim is made for completeness and further details can be obtained in standard texts [3, 28].

Vectors

A vector has (such as force, velocity, angular momentum) magnitude and direction as opposed to a scalar quantity (mass, length, time) which lacks direction.

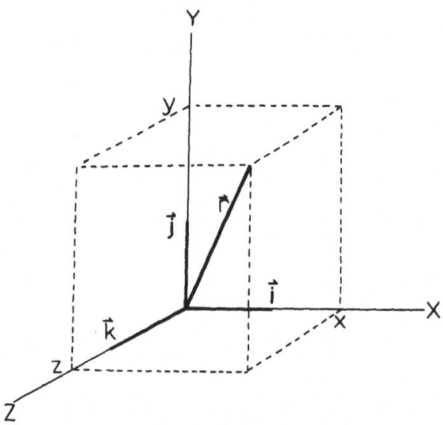

Fig. A 1. Co-ordinate system (XYZ) and unit vectors $\vec{i}, \vec{j}, \vec{k}$ defining the vector $\vec{r}$

A vector $\vec{r}$ in the three dimensional Cartesian co-ordinate system (XYZ) may be defined in terms of three scalar quantities (xyz) and three vectors $\vec{i}\,\vec{j}\,\vec{k}$ with unit magnitude directed along the Cartesian axes (XYZ) respectively.

$$\vec{r} = x\vec{i} + y\vec{j} + z\vec{k}$$

A second vector r' may be defined by three more scalar quantities (x'y'z')

$$\vec{r}' = x'\vec{i} + y'\vec{j} + z'\vec{k}.$$

The *scalar product* of two vectors $\vec{r}.\vec{r}'$ is given by

$$\vec{r}.\vec{r}' = xx' + yy' + zz'$$

since $\vec{i}\,\vec{j}$ and $\vec{k}$, by definition, are mutually orthogonal. The two vectors r and r' are said to be *orthogonal* when their scalar product is zero.

In more general terms we may consider n scalar quantities (numbers a, b, c, d..n) as defining a vector $\vec{v}_n$ in n-dimensional space. The *orthogonality condition* can then be generalised to include two vectors $\vec{v}_n$ and $\vec{v}_n'$ in n-dimensional space: —

$$\vec{v}_n \cdot \vec{v}_n = aa' + bb' + cc' + \ldots + nn' = 0$$

The notation used to define a vector is often of a row or column type: —

$$[a, b, c, \ldots] \quad \text{row vector}$$

$$\begin{bmatrix} a \\ b \\ c \\ \vdots \end{bmatrix} \quad \text{column vector}$$

This is a special case of matrix notation.

Matrices

It is very useful to be able to express the effects of symmetry operations or permutations in terms of a matrix notation which may be used to define the properties of the appropriate mathematical group in a mathematically more rigorous way (see Appendix B).

Matrix notation is treated here specifically in the context of the effects of symmetry operations and is defined by examples below.

Matrix Notation for a Reflection

A vector $[x_1, y_1]$ lying in the XY plane of a co-ordinate system XYZ is transformed into $[x_1, -y_1]$ by reflection in the XZ plane and $[-x_1, y_1]$ by reflection in the YZ plane.

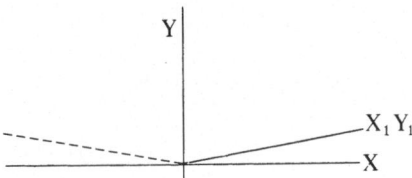

These transformations can be expressed alternatively as: —

$$\sigma_{XZ} x_1 = x_1, \quad \sigma_{XZ} y_1 = -y_1$$
$$\sigma_{YZ} x_1 = -x_1, \quad \sigma_{YZ} y_1 = y_1$$

or

$$\sigma_{XZ} \begin{bmatrix} x_1 \\ y_1 \end{bmatrix} = \begin{bmatrix} 1 & 0 \\ 0 & -1 \end{bmatrix} \begin{bmatrix} x_1 \\ y_1 \end{bmatrix} = \begin{bmatrix} x_1 \\ -y_1 \end{bmatrix}$$

$$\sigma_{YZ} \begin{bmatrix} x_1 \\ y_1 \end{bmatrix} = \begin{bmatrix} -1 & 0 \\ 0 & 1 \end{bmatrix} \begin{bmatrix} x_1 \\ y_1 \end{bmatrix} = \begin{bmatrix} -x_1 \\ y_1 \end{bmatrix}$$

The arrays of numbers (1, 0 and -1 in these cases) in square brackets are the matrix notations for the operations and can be seen to be the coefficients of the co-ordinates in the transformations:

$$\sigma_{XZ} \qquad\qquad\qquad\qquad \sigma_{YZ}$$

$$\begin{bmatrix} x_1 \\ y_1 \end{bmatrix} \rightarrow \begin{bmatrix} 1x_1 + & 0y_1 \\ 0x_1 + & -1y_1 \end{bmatrix} : \begin{bmatrix} x_1 \\ x_1 \end{bmatrix} \rightarrow \begin{bmatrix} -1x_1 + 0y_1 \\ 0x_1 + 1y_1 \end{bmatrix}$$

The matrices representing the operations are composed of two rows (i, horizontal) and two columns (j, vertical) of *elements* a_{ij} which define *the order* as 2×2. In general the order of the matrix is denoted by $m \times n$ and m need not be equal to n. However in all the cases given here m = n and the matrices are called *square matrices*. The operation called the *identity operation* which leaves any system as it was is represented by a *unit matrix* where all of the elements except the *principal or diagonal* elements ($= 1$) are zero.

$$\varepsilon \equiv \begin{bmatrix} 1 & 0 \\ 0 & 1 \end{bmatrix} \text{ in this case.}$$

The vectors have been denoted by single column matrices for reasons which will become apparent when *matrix multiplication* is considered below.

Matrix Notation for a Rotation

A vector $\vec{r_1}$ $(x_1 y_1)$ lying in the XY plane of a co-ordinate system XYZ is transformed into a second vector $\vec{r_2}$ $(x_2 y_2)$ by a rotation through an angle θ about the origin

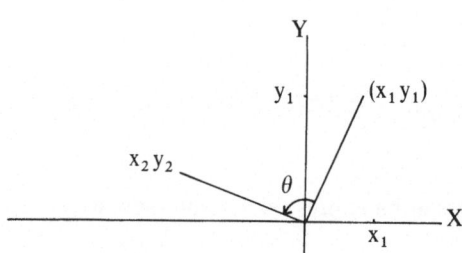

The transformation can be made by examining the behaviour of the x and y components of $\vec{r_1}$.

The vector $\vec{i}x_1$ is transformed into a vector $\vec{r}'$ with components $-\vec{i}x_1 \cos\theta$ and $\vec{j}x_1 \sin\theta$ and the vector $\vec{j}y_1$ becomes a vector $\vec{r}''$ with components $-\vec{i}y_1 \sin\theta$, $-\vec{j}y_1 \cos\theta$. These may be expressed in matrix

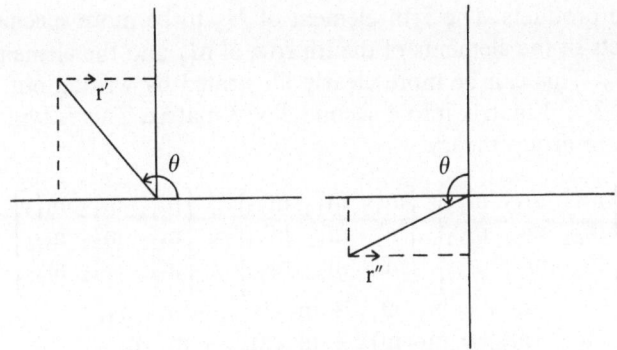

notation as

$$\theta \begin{bmatrix} x_1 \\ 0 \end{bmatrix} = \begin{bmatrix} -x_1 \cos \theta \\ x_1 \sin \theta \end{bmatrix}$$

$$\theta \begin{bmatrix} 0 \\ y_1 \end{bmatrix} = \begin{bmatrix} -y_1 \sin \theta \\ -y_1 \cos \theta \end{bmatrix}$$

$$\begin{bmatrix} x_1 \\ y_1 \end{bmatrix} = \begin{bmatrix} x_1 \\ 0 \end{bmatrix} + \begin{bmatrix} 0 \\ y_1 \end{bmatrix}$$ using the simple property of matrix addition.

Therefore

$$\theta \begin{bmatrix} x_1 \\ y_1 \end{bmatrix} = \begin{bmatrix} -x_1 \cos \theta - y_1 \sin \theta \\ + x_1 \sin \theta - y_1 \cos \theta \end{bmatrix}$$

and θ can be expressed in matrix notation as

$$\theta \equiv \begin{bmatrix} -\cos \theta, & -\sin \theta \\ +\sin \theta, & -\cos \theta \end{bmatrix}$$

Reflection of $\vec{r_2}$ in the YZ plane generates a third vector $\vec{r_3}$

$$\vec{r_3} = \begin{bmatrix} x_1 \cos \theta + y_1 \sin \theta \\ x_1 \sin \theta - y_1 \cos \theta \end{bmatrix} = \begin{bmatrix} x_3 \\ y_3 \end{bmatrix}$$

Matrix Multiplication

The multiplication of two matrices M_1 and M_2 can only be accomplished if the number of columns in M_1 is equal to the number of rows in M_2. Thus if M_1 is of order $(r_1 \times c_1)$ while M_2 is of order $(r_2 \times c_2)$ and $c_1 = r_2$ then M_1 and M_2 are *conformable*. The result of the product of $M_1 \times M_2 = M_3$ will be of order $(r_1 \times c_2)$. The mechanics of multiplication may be described as follows. The elements of the product matrix M_3 are obtained by multiplying the elements of the rows of M_1 into corresponding elements of the columns of M_2 and summing over all these products. The ijth element of M_3 to be more specific, is the sum of the products of the elements of the ith row of M_1 and the elements of the jth column of M_2. This can be more clearly illustrated by writing out the result of multiplying a 3×3 matrix into a second 3×3 matrix. This is typical of matrix manipulation in group theory.

$$\begin{bmatrix} m_{11} & m_{12} & m_{13} \\ m_{21} & m_{22} & m_{23} \\ m_{31} & m_{32} & m_{33} \end{bmatrix} \begin{bmatrix} m'_{11} & m'_{12} & m'_{13} \\ m'_{12} & m'_{22} & m'_{23} \\ m'_{31} & m'_{32} & m'_{33} \end{bmatrix} = \begin{bmatrix} m''_{11} & m''_{12} & m''_{13} \\ m''_{21} & m''_{22} & m''_{23} \\ m''_{31} & m''_{32} & m''_{33} \end{bmatrix}$$

$$m''_{11} = m_{11}m'_{11} + m_{12}m'_{21} + m_{13}m'_{31}$$
$$m''_{12} = m_{11}m'_{12} + m_{12}m'_{22} + m_{13}m'_{32}$$
$$m''_{13} = m_{11}m'_{13} + m_{12}m'_{23} + m_{13}m'_{33}$$
$$m''_{21} = m_{21}m'_{11} + m_{22}m'_{21} + m_{23}m'_{31}$$
$$m''_{22} = m_{21}m'_{12} + m_{22}m'_{22} + m_{23}m'_{32}$$
$$m''_{23} = m_{21}m'_{13} + m_{22}m'_{23} + m_{23}m'_{33}$$
$$m''_{31} = m_{31}m'_{12} + m_{32}m'_{22} + m_{33}m'_{31}$$
$$m''_{32} = m_{31}m'_{11} + m_{32}m'_{22} + m_{33}m'_{32}$$
$$m''_{33} = m_{31}m'_{13} + m_{32}m'_{23} + m_{33}m'_{33}$$

In general $M_1 M_2 \neq M_2 M_1$ or matrix multiplication is not *commutative*, but it does, however, always obey the *associative* law

$$M_1(M_2 \times M_3) = (M_1 \times M_2)M_3,$$

Inverse of a Matrix

The inverse of a matrix M_1 is, by definition, a matrix denoted by M^{-1} such that $MM^{-1} = M^{-1}M =$ the unit matrix, ε.

In the context of the matrix representation of symmetry operations or permutations (P), M^{-1} is the matrix representation corresponding to the operation P^{-1} which restores the system exactly to the original configuration ($PP^{-1} = \varepsilon$).

It can be shown that only matrices with non-vanishing determinants can have inverses and since only square determinants can be non zero, it follows that only square matrices can have inverses.

The *determinant* of a matrix

$$\begin{bmatrix} m_{11} & m_{12} \\ m_{21} & m_{22} \end{bmatrix} \quad \text{is} \quad \begin{vmatrix} m_{11} & m_{12} \\ m_{21} & m_{22} \end{vmatrix}$$

which expanded gives $|m_{11}m_{22} - m_{21}m_{12}|$. The expansion is carried out by multiplying a chosen element into the sub-determinant created by eliminating the row and column containing the chosen element, alternating the signs of the columns so that odd numbered columns, (1,3 etc.) have a positive sign and even columns (2,4 etc.) have a negative sign. Thus the determinant

$$\begin{vmatrix} m_{11} & m_{12} & m_{13} \\ m_{21} & m_{22} & m_{23} \\ m_{31} & m_{32} & m_{33} \end{vmatrix} = 0$$

can be expanded as

$$m_{11}[m_{22}m_{33} - m_{23}m_{32}] - m_{12}[m_{21}m_{33} - m_{23}m_{31}] + m_{13}[m_{21}m_{32} - m_{22}m_{31}].$$

A matrix M having a determinant $|D|$ which equals zero is said to be *singular*. All operations considered later have corresponding inverses so that only *non-singular* matrices occur in appendix B.

Character of Square Matrix

The sum of the diagonal elements of a square matrix is called the *character*, χ of the matrix. This has a special importance in characterising the matrix, providing an abbreviated notation with a specific meaning in symmetry groups which can be conveniently handled in group theoretical formulae.

$$\text{Character of } M = \chi_M = \sum_i m_{ii}$$

Conjugate Matrices; The Definition of a Class

Two matrices M_1 and M_2 are *conjugate* with one another if there exists a third matrix M_3 with inverse M_3^{-1} such that −

$$M_1 = M_3^{-1} M_2 M_3$$

This relation can be expressed in words by saying that M_1 is the *similarity transform* of M_2 by M_3.

A set of matrices (or symmetry operations) which generate one another by similarity transformations of the type shown above are conjugate with one another and constitute a *class* of a mathematical group.

Summary

Matrix notation and some of the properties of matrices have been introduced with the symmetry group in mind. Some overlap into the theory of groups has been introduced in discussing characters and classes of matrices. The theory of groups is expanded a little more in appendix B once again allied to the symmetry group in particular.

Appendix B

The Symmetry Group, Character Table, Determinant Factorisation and Symmetrised Wavefunctions for the $A_2A_2'XX'X''X'''$ System

1. The Model for the Non-rigid $A_2A_2'XX'X''X'''$ System

This appendix is centred around the NMR group of the non-rigid system typified by α,α' dichloro-p-xylene at temperatures where the two chloromethyl groups may be considered to spend periods of time in the possible conformational sites small compared with the reciprocal of the methylene proton-ortho ring proton coupling constant [2].

The Hamiltonian of the system can be defined using this non-rigid model in terms of two Larmor frequencies and six coupling constants indicated in Fig. B 1, summarised in the NMR notation, $A_2A_2'XX'X''X'''$.

Fig. B 1. Model for the non-rigid $A_2A_2'XX'X''X'''$ system characterised by the scalar coupling constants

The NMR symmetry group can then be defined as consisting of permutations of identical nuclei which do not alter the NMR parameters. In other words, the permutations of the group leave the Hamiltonian unchanged or the Hamiltonian is *invariant* under the operations of the group. The criterion of feasibility must also be invoked so that no bonds are broken or insuperable energy barriers disregarded in carrying out the operations of the group.

The approach here is empirical to the extent that the allowed permutations of nuclei are accumulated by trial and error with no absolute certainty that all the operations of the group have been recorded. However having completed the empirical survey, it is possible to qualify the set of operations obtained, in a quantitative way, by testing to see if the laws of the mathematical group are obeyed. This will be done after the empirical set of operations has been derived.

2. Empirical Permutation Group of the $A_2A_2'XX'X''X'''$ NMR System

The permutations

    (12)   exchange 1 and 2
    (34)   exchange 3 and 4
  (12)(34)   exchange 1 and 2, and 3 and 4, at the same time,

are feasible since the system has been defined by saying that there is rotation about the methylene-carbon-ring-carbon bonds. Similarly (56)(78) is allowed since the original configuration can be restored by a rotation of the benzene ring relative to the two methylene groups. Variations on this theme lead to three more similar permutations,

$$(12)(56)(78), \ (34)(56)(78) \text{ and } (12)(34)(56)(78).$$

Eight additional permutations can be isolated which can be annulled by rotations about an axis perpendicular to the 'para-axis' of the molecule (C_2X) in Fig. B 2.

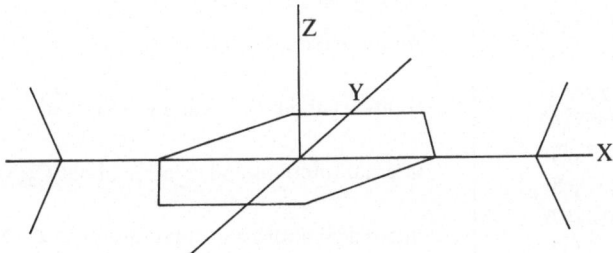

Fig. B 2. Co-ordinate system defining the rotations of the molecule which annull feasible permutations

These are,

    (13)(24)(57)(68),   (14)(23)(57)(68)
    (13)(24)(58)(67),   (14)(23)(58)(67)
    (1324)(57)(68),   (1423)(57)(68)
    (1324)(58)(67),   (1423)(58)(67).

The permutations of the group are summarised in Table B 1 and characterised by the molecular motions defining their feasibility.

3. The Mathematical Group

It is possible to check the NMR permutation group defined above for the $A_2A_2'XX'X''X'''$ system using the laws and theorems obeyed by mathematical groups. These are summarised here and applied immediately afterwards to verify the completeness of Table B 1.

3.1. Properties of the Mathematical Group [3]

The group is defined here in terms of '*elements*' which in the context of symmetry groups may be symmetry operations or permutations, or the appropriate matrix representation of these.

1. The result of combining two elements A and B (written A × B or B × A where A may be the same as B) must be an element in the group.

Table B 1. *Feasible permutations of the nuclei of the $A_2A_2'XX'X''X'''$ system*

Permutation		Restoring 'motion'[a]
P_1	E (identity)	leave the system alone
P_2	(12)	
P_3	(34)	
P_4	(12)(34)	internal rotations
P_5	(56)(78)	
P_6	(12)(56)(78)	
P_7	(34)(56)(78)	
P_8	(12)(34)(56)(78)	rotation about $C_2(X)$
P_9	(13)(24)(57)(68)	rotation about $C_2(Y)$
P_{10}	(14)(23)(57)(68)	
P_{11}	(13)(24)(58)(67)	rotations about $C_2(Y)$ and $C_2(X)$
P_{12}	(14)(23)(58)(67)	
P_{13}	(1324)(57)(68)	internal rotations and rotation about $C_2(Y)$
P_{14}	(1423)(57)(68)	
P_{15}	(1324)(58)(67)	internal rotations and rotations about $C_2(Y)$ and $C_2(X)$
P_{16}	(1423)(58)(67)	

[a] The form of the Hamiltonian allows the replacement of the chloromethyl group by a planar CH_2 group, since the chlorine nucleus is quadrupolar relaxed and does not contribute.

In general A × B ≠ B × A, that is to say the commutative law does not in general hold, so that AB = C and BA = D. In those cases where C = D for all combinations, combination is commutative and the group is called an *Abelian* group.

2. The group must include the identity element E which commutes with all other elements and leaves them unchanged, thus; EX = XE = X.

3. The associative law of multiplication must hold and can be illustrated by

$$A(BC) = (AB)C.$$

This means that if B is first 'combined' with C(B × C) and the result 'left' combined with A (A × (B × C)), the element generated is identical with that produced by 'combining' A with B (A × B) and 'right' combining C, (A × B) × C.

4. Every element, X, must have a reciprocal, X^{-1}, which is also an element of the group.

$$XX^{-1} = E = X^{-1}X, \quad E^{-1} = E$$

It can be shown that $(AB..)^{-1} = ... B^{-1}A^{-1}$.

3.2. Mathematical (Permutation) Group of the $A_2A_2'XX'X''X'''$ NMR System

The elements (now permutations) of the group have been derived in an empirical way in section B 2 and these can now be tested using the above rules to verify that the group is complete. Rule 1 applied to the permutations of Table B 1 shows that indeed the permutations in Table B 1 do constitute a complete mathematical set of *order h = 16*!

This is a convenient time to introduce the idea of a *sub-group*. Any mathematical group may be composed of a number of sub-groups containing elements which themselves obey the rules laid out above. The order of any sub-group which exists must be a divisor of h. Thus the permutations E, (12)(34), (12)(34) constitute a sub-group of order 4 in Table B 1. Similarly the first eight permutations in Table B 1 constitute a sub-group of order 8.

The permutations of the group can also be separated into smaller sets, called classes, using the similarity transformation operation introduced for the matrix representation in appendix A. A complete set of permutations which are conjugate to one another is called a class of the group. Remember that two permutations P_i and P_j are said to be conjugate when there exists a third permutation P_k such that

$$P_i = P_k^{-1}P_jP_k.$$

P_i is the similarity transform of P_j by P_k.

The classes of permutations in Table B 1 can be found using the similarity transformation method and they are shown in Table B 2.

Table B 2. *Classes of permutations for the non-rigid*
$A_2A_2'XX'X''X'''$ system

E	(12)(34)(56)(78)
(12), (34)	(12)(24)(57)(68),(14)(23)(57)(68)
(12)(34)	(13)(24)(58)(67),(14)(23)(58)(67)
(56)(78)	(1324)(57)(68),(1423)(57)(68)
(12)(56)(78),(34)(56)(78)	(1324)(58)(67),(1423)(58)(67)

There are therefore ten classes of the group with order either one or two.
Theorem. The orders of all classes are integral factors of the order of the group.

It will be shown that the number of classes is of fundamental importance in deriving the mathematical representation of the group.

3.3. Mathematical Representation of the Group

The idea of the matrix representation of a symmetry element has been introduced and illustrated in appendix A. *A representation of a group appropriate for discussion here may be defined as a set of matrices each corresponding to each single permutation in the group, which can be combined among themselves in a way analogous to the way the group elements — in this case, permutations, — combine.*

Any set of algebraic functions or vectors may be used as the *basis* for a representation of the group. In order to use algebraic functions as a basis, they are considered to be the components of a vector and the matrices which show that vector is transformed by each permutation are determined.

In this case it seems appropriate to use internal functions [4] since the system is non-rigid and for that reason inter-nuclei vectors have been adopted as a basis for the permutation group of the $A_2A_2'XX'X''X'''$ system. These are given in Table B 3 and the effect of the permutations of the group are illustrated.

Table B 3 shows that the vectors chosen as the basis for the mathematical representation can in fact be divided into five sets:

$$\frac{\vec{12}}{\vec{34}}; \quad \frac{\vec{13}}{\vec{24}}; \quad \frac{\vec{56}}{\vec{78}}; \quad \frac{\vec{57}}{\vec{68}} \quad \frac{\vec{58}}{\vec{67}}; \quad \frac{\vec{14}}{\vec{23}}$$

Table B 3. *Vector used as a basis for the representation of the group of $A_2A_2'XX'X''X'''$ systems*

	$\vec{12}$	$\vec{34}$	$\vec{13}$	$\vec{24}$	$\vec{14}$	$\vec{23}$	$\vec{56}$	$\vec{78}$	$\vec{57}$	$\vec{68}$	$\vec{58}$	$\vec{67}$
E												
(12)	$-\vec{12}$	$\vec{34}$	$\vec{23}$	$\vec{14}$	$\vec{24}$	$\vec{13}$	$\vec{56}$	$\vec{78}$	$\vec{57}$	$\vec{68}$	$\vec{58}$	$\vec{67}$
(34)	$\vec{12}$	$-\vec{34}$	$\vec{14}$	$\vec{23}$	$\vec{13}$	$\vec{24}$	$\vec{56}$	$\vec{78}$	$\vec{57}$	$\vec{68}$	$\vec{58}$	$\vec{67}$
(12)(34)	$-\vec{12}$	$-\vec{34}$	$\vec{24}$	$\vec{13}$	$\vec{23}$	$\vec{14}$	$\vec{56}$	$\vec{78}$	$\vec{57}$	$\vec{68}$	$\vec{58}$	$\vec{67}$
(56)(78)	$\vec{12}$	$\vec{34}$	$\vec{13}$	$\vec{24}$	$\vec{14}$	$\vec{23}$	$-\vec{56}$	$-\vec{78}$	$\vec{68}$	$\vec{57}$	$\vec{67}$	$\vec{58}$
(12)(56)(78)	$-\vec{12}$	$\vec{34}$	$\vec{23}$	$\vec{14}$	$\vec{24}$	$\vec{13}$	$-\vec{56}$	$-\vec{78}$	$\vec{68}$	$\vec{57}$	$\vec{67}$	$\vec{58}$
(34)(56)(78)	$\vec{12}$	$-\vec{34}$	$\vec{14}$	$\vec{23}$	$\vec{13}$	$\vec{24}$	$-\vec{56}$	$-\vec{78}$	$\vec{68}$	$\vec{57}$	$\vec{67}$	$\vec{58}$
(12)(34)(56)(78)	$-\vec{12}$	$-\vec{34}$	$\vec{24}$	$\vec{13}$	$\vec{23}$	$\vec{14}$	$-\vec{56}$	$-\vec{78}$	$\vec{68}$	$\vec{57}$	$\vec{67}$	$\vec{58}$
(13)(24)(57)(68)	$\vec{34}$	$\vec{12}$	$-\vec{13}$	$-\vec{24}$	$-\vec{23}$	$-\vec{14}$	$\vec{78}$	$\vec{56}$	$-\vec{57}$	$-\vec{68}$	$-\vec{67}$	$-\vec{58}$
(14)(23)(57)(68)	$-\vec{34}$	$-\vec{12}$	$-\vec{24}$	$-\vec{13}$	$-\vec{14}$	$-\vec{23}$	$\vec{78}$	$\vec{56}$	$-\vec{57}$	$-\vec{68}$	$-\vec{67}$	$-\vec{58}$
(13)(24)(58)(67)	$\vec{34}$	$\vec{12}$	$-\vec{13}$	$-\vec{24}$	$-\vec{23}$	$-\vec{14}$	$-\vec{78}$	$-\vec{56}$	$-\vec{68}$	$-\vec{57}$	$-\vec{58}$	$-\vec{67}$
(14)(23)(58)(67)	$-\vec{34}$	$-\vec{12}$	$-\vec{24}$	$-\vec{13}$	$-\vec{14}$	$-\vec{23}$	$-\vec{78}$	$-\vec{56}$	$-\vec{68}$	$-\vec{57}$	$-\vec{58}$	$-\vec{67}$
(1423)(57)(68)	$-\vec{34}$	$\vec{12}$	$-\vec{14}$	$-\vec{23}$	$-\vec{24}$	$-\vec{13}$	$\vec{78}$	$\vec{56}$	$-\vec{57}$	$-\vec{68}$	$-\vec{67}$	$-\vec{58}$
(1324)(57)(68)	$\vec{34}$	$-\vec{12}$	$-\vec{23}$	$-\vec{14}$	$-\vec{13}$	$-\vec{24}$	$\vec{78}$	$\vec{56}$	$-\vec{57}$	$-\vec{68}$	$-\vec{67}$	$-\vec{58}$
(1423)(58)(67)	$-\vec{34}$	$\vec{12}$	$-\vec{14}$	$-\vec{23}$	$-\vec{24}$	$-\vec{13}$	$-\vec{78}$	$-\vec{56}$	$-\vec{68}$	$-\vec{57}$	$-\vec{58}$	$-\vec{67}$
(1324)(58)(67)	$\vec{34}$	$-\vec{12}$	$-\vec{24}$	$-\vec{13}$	$-\vec{14}$	$-\vec{23}$	$-\vec{78}$	$-\vec{56}$	$-\vec{68}$	$-\vec{57}$	$-\vec{58}$	$-\vec{67}$

The vectors in each set are permuted into one another by the permutations of the group often with a minus sign indicating a reversal of the vector. The matrix representations corresponding to the classes of permutations of the group for each set of vectors can be derived from Table B 3. For example the permutation (12) acting on the set of vectors $\vec{12}$, $\vec{34}$ can be represented by

$$(12) \begin{bmatrix} \vec{12} \\ \vec{34} \end{bmatrix} = \begin{bmatrix} -1 & 0 \\ 0 & 1 \end{bmatrix} \begin{bmatrix} \vec{12} \\ \vec{34} \end{bmatrix} = \begin{bmatrix} -\vec{12} \\ \vec{34} \end{bmatrix}$$

using the matrix notation.
Also

$$(34) \begin{bmatrix} \vec{12} \\ \vec{34} \end{bmatrix} = \begin{bmatrix} 1 & 0 \\ 0 & -1 \end{bmatrix} \begin{bmatrix} \vec{12} \\ \vec{34} \end{bmatrix} = \begin{bmatrix} \vec{12} \\ -\vec{34} \end{bmatrix}$$

so that the characters of the matrix representations of (12) and (34), taken as the sum of the diagonal matrix elements, are the same and equal to zero.

An example of the transformation matrix of the four vector set is given by

$$(13)(24)(57)(68) \begin{bmatrix} \vec{13} \\ \vec{24} \\ \vec{14} \\ \vec{23} \end{bmatrix} = \begin{bmatrix} -1 & 0 & 0 & 0 \\ 0 & -1 & 0 & 0 \\ 0 & 0 & 0 & -1 \\ 0 & 0 & -1 & 0 \end{bmatrix} \begin{bmatrix} \vec{13} \\ \vec{24} \\ \vec{14} \\ \vec{23} \end{bmatrix}$$

where the character χ, is -2.

The characters of the matrix representations of the transformation behaviour of the five sets of vectors are given in Table B 4.

Table B 4. *Characters χ of matrix representations of the classes of permutations of the group acting on the vectors which form a basis for representations Γ of the group*

P_i	$\Gamma(I)$ $\vec{12}$ $\vec{34}$	$\Gamma(II)$ $\vec{13}$ $\vec{24}$ $\vec{14}$ $\vec{23}$	$\Gamma(III)$ $\vec{56}$ $\vec{78}$	$\Gamma(IV)$ $\vec{57}$ $\vec{68}$	$\Gamma(V)$ $\vec{58}$ $\vec{67}$	Class notation
E	2	4	2	2	2	C_1
(12)	0	0	2	2	2	C_2
(12)(34)	-2	0	-2	2	2	C_3
(56)(78)	2	4	-2	0	0	C_4
(12)(56)(78)	0	0	-2	0	0	C_5
(12)(34)(56)(78)	-2	0	0	0	0	C_6
(13)(24)(56)(67)	0	-2	0	0	-2	C_7
(13)(24)(57)(68)	0	-2	0	-2	0	C_8
(1423)(57)(68)	0	0	0	-2	0	C_9
(1423)(58)(67)	0	0	0	0	-2	C_{10}

Table B 4 illustrates just a few of the many possible basis representations of the group. Others would be vectors such as $\vec{15}$, $\vec{16}$, $\vec{17}$, $\vec{18}$, $\vec{25}$, $\vec{26}$, $\vec{27}$, $\vec{28}$ ... etc. which can be shown to constitute sets in a similar way to that given in Table B 3.

However, for any group, only a limited number of representations are of fundamental significance.

It is possible to distinguish between two types of representations. A *reducible representation* is a set of matrices which can be reduced to combinations of two or more representations of smaller dimensions, (the dimension of a representation 1_i is the order of the square matrices which constitute it) by a similarity transformation involving some matrix M', thus:

$$M'^{-1}MM' = \begin{bmatrix} M_1 & \\ & M_2 \end{bmatrix}$$

An *irreducible representation* cannot be reduced by any similarity transformation and it is this type of representation which is of fundamental importance. There is a number of properties and rules associated with irreducible representations (given here without proof, but see reference 3) which form the foundation for the derivation of the mathematical representation of the group.

The symbols used in the rules given below have the following significance:

1_i dimension of the i-th representation = order of the matrices which constitute it.

P_i permutations of the group.

$\chi_i(P)$ character of the matrix representation of P in the i-th irreducible representation of the group.

Rule 1

The characters of all matrices belonging to operations in the same class are identifical in a given representation. (This is true for reducible and irreducible representations). Table B 3 and subsequent discussion confirm this.

Rule 2

The number of irreducible representations of a group is equal to the number of classes in the group. This leads to the conclusion that there are ten irreducible representations in the mathematical group representation of $A_2A_2'XX'X''X'''$ systems.

Rule 3

The sum of the squares of the dimensions of the irreducible representations of a group is equal to the order of the group.

$$\sum l_i^2 = l_1^2 + l_2^2 + l_3^2 ... = h$$

Also the sum of the squares of the characters of the identity representation E is equal to the order of the group since the character of the representation of E is equal to the order of the representation.

$$\sum_i [\chi_i(E)]^2 = h$$

Rule 4

The sum of the squares of the characters in any irreducible representation equals h,

$$\sum_P [\chi_i(P)]^2 = h$$

Rule 5

The vectors whose components are the characters of two different representations are orthogonal,

$$\sum_P \chi_i(P)\chi_j(P) = 0, \text{ when } i \neq j.$$

These rules can be applied directly to Table B 4 to establish which of the representations,

$$\Gamma(\text{I}),\ \Gamma(\text{II}),\ \Gamma(\text{III}),\ \Gamma(\text{IV}) \text{ and } \Gamma(\text{V}),$$

if any, are irreducible.

$\Gamma(\text{I})$ obeys rule three and rule four

$$\sum [1_i(\text{I})]^2 = 16 \quad \text{and} \quad \sum_i [\chi_i(P)\text{I}]^2 = 16.$$

but $\Gamma(\text{II}), \Gamma(\text{III}), \Gamma(\text{IV})$ and $\Gamma(\text{V})$ do not. $\Gamma(\text{I})$ is therefore adopted as an irreducible representation of the group and labelled here Γ_5 (anticipating the final result!).

$\Gamma(\text{II}), \Gamma(\text{III}), \Gamma(\text{IV})$ and $\Gamma(\text{V})$ are obviously reducible. The matrix representations in $\Gamma(\text{II}), \Gamma(\text{III}), \Gamma(\text{IV})$ and $\Gamma(\text{V})$ have a dimension of two; it follows that they are linear combinations of representations with unit dimensions since the dimension two can only be reduced in one way, $2 = 1 + 1$!

$\Gamma(\text{IV})$ is adopted as an example here and the possible linear combinations can be summarised as indicated below. The C's refer to the ten classes of the group defined in Table B 4.

	C_1	C_2	C_3	C_4	C_5	C_6	C_7	C_8	C_9	C_{10}
	1	1	1	± 1	± 1	± 1	± 1	-1	-1	± 1
	1	1	1	∓ 1	∓ 1	∓ 1	∓ 1	-1	-1	∓ 1
$\Gamma(\text{IV})$	2	2	2	0	0	0	0	-2	-2	0

It is possible to show using rule 5 that only two of the above representations are orthogonal to Γ_5 and to one another. These are:

Γ_2	1	1	1	1	1	1	-1	-1	-1	-1
Γ_3	1	1	1	-1	-1	-1	1	-1	-1	1

Γ_2 and Γ_3 are tentatively adopted as irreducible representations of the group and checked as other new irreducible representations are found. A new irreducible representation is generated by combining Γ_2 and Γ_3 as

$$\Gamma_2 \times \Gamma_3 = \Gamma_4$$

Γ_4	1	1	1	-1	-1	-1	-1	1	1	-1

Attention is now focussed on $\Gamma(\text{II})$ which has a matrix representation with dimension four and must be reducible since it does not satisfy rule 4. It has been pointed out in section 4 that combinations of irreducible representations can be reduced to their component representations using a standard eq. 4(iv). This is given here again, to assist continuity in this appendix, in a slightly different context

$$N^{\Gamma_i} = \frac{1}{h} \sum_c g_c \, \chi(P)_c \, \chi_i(P)_c \ldots \ldots \qquad \text{B(i)}$$

where N^{Γ_i} is the number of times the representation Γ_i occurs in the reducible representation, h is the order of the group, g_c is the order of the class C, $\chi(P)_c$ are the characters of the reducible representation for each permutation P in class C, and $\chi_i(P)$ are the characters of the irreducible representation Γ_i.

In this instance therefore $\chi(P)$ are given by the characters of the reducible representation,

$$\Gamma(\text{II}) \qquad 4 \quad 0 \quad 0 \quad 4 \quad 0 \quad 0 \quad -2 \quad -2 \quad 0 \quad 0,$$

which correspond to the ten classes of the group listed in Table B 2. It is found using eq. B(i) that Γ_3 and Γ_5 occur once each in $\Gamma(\text{II})$ since

$$N^{\Gamma_3} = 1/16[1 \times 4 \times 1 + 2 \times 0 \times 1 + 1 \times 0 \times 1 + 1 \times 4 \times 1 + 2 \times 0 \times 1 + 2 \times 0 \times 1$$
$$+ 2 \times (-2) \times (-1) + 2 \times (-2) \times (-1) + 2 \times 0 \times (-1) + 2 \times 0 \times (-1)] = 1$$

and

$$N^{\Gamma_5} = 1/16 \,[1 \times 4 \times 2 + 2 \times 0 \times 0 + 1 \times 0 \times (-2) + 1 \times 4 \times 2 + 2 \times 0 \times 0$$
$$+ 2 \times 0 \times (-2) + (2 \times (-2) \times 0) + 2 \times (-2) \times 0 + 2 \times 0 \times 0 + 2 \times 0 \times 0] = 1$$

Combining Γ_2 and Γ_5 as $\Gamma_2 + \Gamma_5$ gives

$$\Gamma_2 + \Gamma_5 \qquad 3 \quad 1 \quad -1 \quad 3 \quad 1 \quad -1 \quad -1 \quad -1 \quad -1 \quad -1$$

and $\Gamma(\text{II}) - (\Gamma_2 + \Gamma_5)$ gives Γ_6

$$\Gamma_6 \qquad 1 \quad -1 \quad 1 \quad 1 \quad -1 \quad 1 \quad 1 \quad -1 \quad 1 \quad 1$$

Combining Γ_6 with Γ_2, Γ_3 and Γ_4 gives

$\Gamma_7 = \Gamma_2 \times \Gamma_6$	1	-1	1	1	-1	1	1	1	-1	-1
$\Gamma_8 = \Gamma_3 \times \Gamma_6$	1	-1	1	-1	1	-1	-1	1	-1	1
$\Gamma_9 = \Gamma_4 \times \Gamma_6$	1	-1	1	-1	1	-1	1	-1	1	-1

and a Γ_{10} can be generated as $\Gamma_5 \times \Gamma_9$

$$\Gamma_{10} \qquad 2 \quad 0 \quad -2 \quad -2 \quad 0 \quad 2 \quad 0 \quad 0 \quad 0 \quad 0$$

The complete table of characters of the ten irreducible representations of the group are given in Table B 5.

Table B 5. *Character table for the mathematical group of the* $A_2A_2'XX'X''X'''$ *system*

	E	(12) (34)	(12)(34)	(56)(78)	(12)(56)(78) (34)(56)(78)	(12)(34)(56)(78)	(14)(23)(58)(67) (13)(24)(58)(67)	(14)(23)(57)(68) (13)(24)(57)(68)	(1423)(57)(68) (1324)(57)(68)	(1423)(58)(67) (13)(24)(58)(67)
Γ_1	1	1	1	1	1	1	1	1	1	1
Γ_2	1	1	1	1	1	1	-1	-1	-1	-1
Γ_3	1	1	1	-1	-1	-1	1	-1	-1	1
Γ_4	1	1	1	-1	-1	-1	-1	1	1	-1
Γ_5	2	0	-2	2	0	-2	0	0	0	0
Γ_6	1	-1	1	1	-1	1	-1	-1	1	1
Γ_7	1	-1	1	1	-1	1	1	1	-1	-1
Γ_8	1	-1	1	-1	1	-1	-1	1	-1	1
Γ_9	1	-1	1	-1	1	-1	1	-1	1	-1
Γ_{10}	2	0	-2	-2	0	2	0	0	0	0

The totally symmetric representation Γ_1 can be generated by squaring any one of the other irreducible representations except Γ_5 and Γ_{10}, (i.e. $\Gamma_2 \times \Gamma_2 = \Gamma_1$).

The character table given in Table B 5 is isomorphic with the D_{4h} symmetry group character table. This means that there are the same number of classes (and therefore irreducible representations) and for every symmetry operation in the D_{4h} symmetry group there is a corresponding permutation in the permutation group of the $A_2A_2'XX'X''X'''$ system.

The symmetrised wavefunctions can be constructed as described in section 4.

The number of symmetrised wavefunctions (or degenerate sets of wavefunctions) which transform according to each irreducible representation Γ_i is given by eq. 4 (ii), repeated here –

$$n^{\Gamma_i} = \frac{1}{h} \sum_C g_c \, \chi_c(P) \, \chi_c(P)^{\Gamma_i}$$

where h is the order of the group, g_c the number of permutations in the class C, $\chi_c(P)$ the number of spin product wavefunctions which are not changed by the permutation P in class C and $\chi_c(P)^{\Gamma_i}$ is the character in the irreducible representation Γ_i corresponding to the class C.

The results of this exercise are given in Table B 6.

The wavefunctions themselves can be constructed using eq. 4 (iii)

$$\psi^{\Gamma_i} = \eta \sum_P \chi(P)^{\Gamma_i} \, P \varphi_i$$

where η is a normalising constant, P_{φ_i} a wavefunction generated by the action of the permutation P on a chosen starting wavefunction φ_i and $\chi(P)^{\Gamma_i}$ the character relevant to P in the irreducible representation Γ_i. The symmetrised wavefunctions are given in Table B 7.

Table B 6. *Schematic illustration of the classification of wave-functions for the two groups of $A_2A_2'XX'X''X'''$ systems*

m_T	A nuclei				X nuclei			
	Γ_1	Γ_2	Γ_5	Γ_7	Γ_1	Γ_2	Γ_3	Γ_4
2	1				1			
1	1	1	1		1	1	1	1
0	2	1	1	1	3	1	1	1
−1	1	1	1		1	1	1	1
−2	1				1			

Table B 7. *Symmetrised wavefunctions for the two groups of the $A_2A_2'XX'X''X'''$ system*

m_T	A nuclei	X nuclei
2	0	
1	Γ_1 $(1+2+3+4)$	Γ_1 $(5+6+7+8)$
	Γ_2 $(1+2-3-4)$	Γ_2 $(5+6-7-8)$
	Γ_5 $(1-2)$	Γ_3 $(5-6+7-8)$
		Γ_4 $(5-6-7+8)$
0	Γ_1 $(12+34)$	Γ_1 $(57+68)$
	$(13+14+23+24)$	$(56+78)$
	Γ_2 $(12-34)$	Γ_2 $(58+67)$
	Γ_5 $(13-24)$	Γ_3 $(56-78)$
	Γ_7 $(13+24-14-23)$	Γ_4 $(58-67)$

(i) The constructional details are symmetrical about $m_T = 0$.
(ii) The numbers locate the nuclei with β spin.
(iii) The wavefunctions can be normalised simply by dividing by the sum of the coefficients squared.

The wavefunctions for the complete system can then be constructed simply from products of the two sets and the 'symmetry' of these products is given in Table B 8.

Table B 8. *Classification of products of symmetrised wavefunctions of the two groups of the $A_2A_2'XX'X''X'''$ system*

	$\Gamma_1(7)$	$\Gamma_2(3)$	$\Gamma_3(3)$	$\Gamma_4(3)$	X nuclei
A nuclei					
$\Gamma_1(6)$	$\Gamma_1(42)$	$\Gamma_2(18)$	$\Gamma_3(18)$	$\Gamma_4(18)$	
$\Gamma_2(3)$	$\Gamma_2(21)$	$\Gamma_1(9)$	$\Gamma_4(9)$	$\Gamma_3(9)$	
$\Gamma_5(6)$	$\Gamma_5(42)$	$\Gamma_5(18)$	$\Gamma_{10}(18)$	$\Gamma_{10}(18)$	
$\Gamma_7(1)$	$\Gamma_7(7)$	$\Gamma_6(3)$	$\Gamma_9(3)$	$\Gamma_8(3)$	

The total wavefunction diagram in which the wavefunctions are classified according to the total magnetic quantum number m_T, the permutation symmetry discussed and the magnetic quantum numbers of the component groups (X approximation, see section 5) is identical in form to the secular determinant

$$|H_{ij} - E\,\delta_{ij}| = 0.$$

The word secular essentially implies 'independent of time'. The determinant is formed simply by subtracting the energy term E from all diagonal elements ($\delta_{ij} = 1$ for $i = j$ and 0 for $i \neq j$) and equating the result to zero. The transformation from the matrix notation to the determinant is made by replacing the square brackets by straight lines.

The secular determinant factorises in the manner indicated in Table B 9.

4. Results of Factorisation

The ideas of sub-spectral analysis mentioned in section 5 can be brought to bear on the $A_2A_2'XX'X''X'''$ system. It can be seen in Table B 9 that there are the a_4 and x_4 sub-spectra corresponding to $m_X = \pm 2$ and $m_A = \pm 2$ respectively. These are located by

$$v_a = v_A \pm (J_{AX} + J_{AX''})$$
$$v_x = v_X \pm (J_{AX} + J_{AX''}).$$

In addition it is possible to pick out 1:2:4:2:1 and 2:2:2 sub-patterns within the X part and 1:2:3:2:1 sub-patterns within the A part. These can be identified superficially with aa'bb' sub-spectra in the X part and parts of a_2b_2 sub-spectra in the A part. However it can be shown that the sub-spectral parameters which describe the 1:2:4:2:1 sub-pattern do not correlate completely with those characterising the 2:2:2 part [15]. It is possible to regard the sub-patterns as separate entities characterised, as they are, by specific equal spacings, and intensity distribution. It may be possible to pick out groups of lines corresponding to those sub-patterns in favourable cases but the assignment, a priori, would be difficult if ambiguity exists. There are two 1:2:4:2:1 sub-patterns in singly degenerate irreducible representation in Table B 9 for example and three in doubly degenerate representations.

The results in Table B 9 can be used for a second purpose. This is to correlate simulated spectra for p-dideuterobenzene computed using a programme written for spin 1/2 nuclei with the experimental data. This would arise where there was no "mixed spin" programme available.

The model adopted in Fig. B 1 can be readily modified for this purpose simply by choosing spin wavefunctions for the two deuterons to be the transformations of the spin 1/2 wavefunctions given in Table B 10.

The representation for the two deuterons in p-dideutero benzene can therefore be summarised as

$$\Gamma = 6\Gamma_1 + 3\Gamma_2.$$

Table B 9. *Schematic illustration of the factorisation of the secular determinant of the* $A_2A'_2XX'X''X'''$ *system using (i)* m_T *(ii) permutation 'symmetry' and (iii) X approximation*

m_T	Γ_1	Γ_2	Γ_3	Γ_4	Γ_5
4	1	1 1	1	1	1
3	1	1 2 1	1 2	1 2	2 1
2	111 2 2	1 4 3 1	1 2 3	1 2 3	4 2 2 1
1	1 4 3 1	2 5 2	2 3 2	2 3 2	2 4 2 2
0	1 2 7 2 1	1 3 4 1	3 2 1	3 2 1	1 2 4 4
−1	1 3 4 1	1 2 1	2 1	2 1	1 2 2
−2	2 2 111	1 1	1	1	1 1 4
−3	1 1	1			1 2
−4	1				1

m_T	Γ_6	Γ_7	Γ_8	Γ_9	Γ_{10}
2		1			2
1	1	1	1	1	2 2
0	1 1	111	1	1	2 2
−1	1	1	1	1	2 2
−2		1		1	2

m_A					
	2 1 0 −1 −2	2 1 0 −1 −2	2 1 0 −1 −2	2 1 0 −1 −2	1 0 −1

Table B 10. *Symmetrised spin wavefunctions for two spin 1, A, nuclei in AA'X X'X"X''' systems correlated with spin 1/2 wavefunctions in $A_2A_2'X X'X"X'''$ systems*

Spin 1/2 wavefunctions	Spin 1 wavefunctions
Γ_1 (1+2+3+4)	(10+01)
Γ_2 (1+2−3−4)	(01−10)
Γ_5 (1−2)	(01−01) = 0
Γ_1 (12+34)	(1−1+−11)
Γ_2 (12−34)	(−11−1−1)
Γ_1 (13+14+23+24)	00
Γ_5 (13−24)	(00−00) = 0
Γ_7 (13+24−14−23)	(00+00−00−00) = 0

The new permutation group which could be used to give the same result is isomorphic with the D_2 symmetry point group since the product wavefunctions $\psi(D_2) \times \psi(X_4)$ can belong only to the irreducible representations $\Gamma_1\ \Gamma_2\ \Gamma_3$ and Γ_4 of Table B 5.

	$\Gamma_1(7)$	$\Gamma_2(3)$	$\Gamma_3(3)$	$\Gamma_4(3)$	X nuclei
D nuclei					
$\Gamma_1(6)$	$\Gamma_1(42)$	$\Gamma_2(18)$	$\Gamma_3(18)$	$\Gamma_4(18)$	
$\Gamma_2(3)$	$\Gamma_2(21)$	$\Gamma_1(9)$	$\Gamma_4(9)$	$\Gamma_3(9)$	

The corresponding symmetrised wavefunction (or secular determinant) diagram is given in Table B 11 and can be compared directly with Table B 9. The data output from the computer can be plotted as an energy level diagram with interconnecting transitions within each irreducible representation. It is then possible to ignore those energy levels and transitions within the $\Gamma_5\ \Gamma_6\ \Gamma_7\ \Gamma_8\ \Gamma_9$ and Γ_{10} representations of Table B 9 to isolate the data relevant to the mixed spin programme.

Combined Symmetry-Composite Particle Approach

The application of the composite particle approach with suitable minimal symmetry to the factorisation of the secular determinant has already been described in section 4.2.3. This method can be applied with equal success to the $A_2A_2'XX'X"X'''$ (or $[A_2XX']_2$ [25, 29]) system providing the same factorisation as the full permutation group without recourse to character table derivation.

The composite particle notation is applicable only to the A nuclei and in particular to each group of magnetically equivalent A nuclei, A_2 and A_2'. The spin states in each group can be characterised by $I_T = 1$ and 0 and the representation of these spin states in composite particle notation is;

$$\Gamma(A_2) = T + S \quad \text{and} \quad \Gamma(A'_2) = T' + S'.$$

Table B 11. *Schematic illustration of the factorisation of the secular determinant of the para-dideuterobenzene problem using the irreducible representation notation of Table B 9.*

m_T	Γ_1					Γ_2					Γ_3					Γ_4				
4	1																			
3	1	1				1	1				1					1				
2	111	2	2			1	2	1			1	2				1	2			
1	1	4	3	1		1	4	3	1		1	2	3			1	2	3		
0	1	2	7	2	1		2	5	2			2	3	2			2	3	2	
−1		1	3	4	1		1	3	4	1			3	2	1			3	2	1
−2			2	2	111			1	2	1				2	1				2	1
−3				1	1				1	1					1					1
−4					1															
m_D	2	1	0	−1	−2	2	1	0	−1	−2	2	1	0	−1	−2	2	1	0	−1	−2

The combinations of these are TT′, SS′, TS′ and ST′. The TT′, and SS′ states do not destroy the 'symmetry' of the spin system which is D_2 as in para-dideuterobenzene. It is therefore possible to form symmetric and anti-symmetric combinations, with respect to the C_2 axis, which permutes the A and A′ nuclei, of the TT′ and SS′ products. These are given in Table B 12 and may be summarised as,

$$\Gamma(TT') = 6A + 3B_1, \quad \Gamma(SS') = 1A.$$

Table B 12. *Symmetric and antisymmetric combinations of the TT′ and SS′ products of $A_2A'_2$ functions in the $A_2A'_2XX'X''X'''$ system*

	Symmetric $(A)^+$	Antisymmetric $(B1)$
TT′	$(1,1)(1,1)'$	
	$(1,1)(1,0)' + (1,0)(1,1)'$	$(1,1)(1,0)' - (1,0)(1,1)'$
	$(1,0)(1,0)$	
	$(1,1)(1,-1) + (1,-1)(1,1)$	$(1,1)(1,-1)' - (1,-1)(1,1)'$
SS′	$(0,0)(0,0)$	

The wavefunctions in this table are characterised by I_T and m_A, $(I_T, m_A)(I_T, m_A)'$. Not all the wavefunctions are shown, but the others can be generated easily for $m_A + m_{A'} = -1$ and -2.

The representation of the four X nuclei in the D_2 symmetry group has been shown to be

$$\Gamma(X_4) = 7A + 3B_1 + 3B_2 + 3B_3$$

The combination of this representation with the TT′ symmetrised wavefunctions shows a one to one correspondence with the $\Gamma_1, \Gamma_2, \Gamma_3$ and Γ_4 representation set out in Table B 9. Similarly the combination of the X_4 wavefunctions with the SS′ state of the A nuclei can be identified with the $\Gamma_6, \Gamma_7, \Gamma_8$ and Γ_9 representation of Table B 9.

It is not possible to use the D_2 symmetry group for the X nuclei in combination with the unsymmetrical TS and ST functions of the A nuclei. Two approaches are valid [34]:

1. The TS′ and ST′ functions can be combined as (TS′ + ST′) and (TS′ − ST′) in the A and B_1 symmetry representations respectively, with the X nuclei in the full D_2 symmetry classification,

$$(3A + 3B_1)(7A + 3B_1 + 3B_2 + 3B_3) = 30A + 30B_1 + 18B_2 + 18B_3$$

This method of approach does not reflect the useful relationship that ST′ and TS′ functions are degenerate [34].

2. The presence of the A_2 and A'_2 groups in TS′ or ST′ states in a molecule reduces the 'effective symmetry' to C_2 and the X functions in this group are classified simply as $10A$ and $6B$.

$$(3\,TS' + 3\,ST')(10A + 6B) = 30(TS')A + 30(ST')A + 18(TS')B + 18(ST')B$$

These two approaches give the same degree of factorisation $(30 + 30 + 18 + 18)$, though the actual basis functions are different, which can be identified with the $\Gamma_5(2 \times 30)$ and $\Gamma_{10}(2 \times 18)$ representations of Table B 9. However it must be pointed out that the classification of the wavefunctions by the composite particle method and the symmetry group are mutually exclusive procedures [1].

The fact that the same results have been obtained using composite particle method with the minimal symmetry in a fraction of the space necessary using symmetry alone is a powerful argument for total spin-symmetry combination!

Appendix C

Group Character Tables

The tables of characters of irreducible representations reproduced here were computed many years ago and can be found in many standard texts. The presentation given by WILSON, DECIUS and CROSS [35] is particularly suitable for calculations in vibrational spectroscopy and SANDORFY [36] has adopted a format relevant to problems in electronic spectroscopy. Either can be adapted for use in nuclear magnetic resonance spectroscopy. The notation in the tables requires a minimum of comment.

A is used to denote a one-dimensional representation if the character is $+1$ for a rotation about the principal axis. B is used if the character is -1, except in cubic groups. E denotes a two dimensional representation and T (sometimes F) indicates a three-dimensional representation. Indices are used to distinguish between representations of the same type — A_1, A_2 etc. A_1 is always the totally symmetric representation. A primed notation is used to distinguish between representations of a given type with different characters ($+1$ or -1) with respect to reflection at a horizontal plane of symmetry thus $A'(+1)$, $A''(-1)$ in C_3 as below. The indices g and u are used according to whether we have $+1$ or -1 with respect to inversion through a centre of symmetry.

Table C 1. *Character tables of C_s, C_i, and the cyclic groups C_n ($n = 2,3,4,5,6$)*

C_s	E	σ_h
A'	1	1
A''	1	-1

C_i	E	i
A_g	1	1
A_u	1	-1

C_2	E	C_2
A	1	1
B	1	-1

C_3	E	C_3	C_3^2	$\varepsilon = e^{2\pi i/3}$
A	1	1	1	
E	$\begin{cases} 1 \\ 1 \end{cases}$	$\begin{matrix} \varepsilon \\ \varepsilon^* \end{matrix}$	$\begin{matrix} \varepsilon^* \\ \varepsilon \end{matrix}$	

C_4	E	C_4	C_2	C_4^3
A	1	1	1	1
B	1	-1	1	-1
E	$\begin{cases} 1 \\ 1 \end{cases}$	$\begin{matrix} i \\ -i \end{matrix}$	$\begin{matrix} -1 \\ -1 \end{matrix}$	$\begin{matrix} -i \\ i \end{matrix}$

C_5	E	C_5	C_5^2	C_5^3	C_5^4	$\varepsilon = e^{2\pi i/5}$
A	1	1	1	1	1	
E_1	$\begin{cases} 1 \\ 1 \end{cases}$	$\begin{matrix} \varepsilon \\ \varepsilon^* \end{matrix}$	$\begin{matrix} \varepsilon^2 \\ \varepsilon^{2*} \end{matrix}$	$\begin{matrix} \varepsilon^{2*} \\ \varepsilon^2 \end{matrix}$	$\begin{matrix} \varepsilon^* \\ \varepsilon \end{matrix}$	
E_2	$\begin{cases} 1 \\ 1 \end{cases}$	$\begin{matrix} \varepsilon^2 \\ \varepsilon^{2*} \end{matrix}$	$\begin{matrix} \varepsilon^* \\ \varepsilon \end{matrix}$	$\begin{matrix} \varepsilon \\ \varepsilon^* \end{matrix}$	$\begin{matrix} \varepsilon^{2*} \\ \varepsilon^2 \end{matrix}$	

C_6	E	C_6	C_3	C_2	C_3^2	C_6^5	$\varepsilon = e^{2\pi i/6}$
A	1	1	1	1	1	1	
B	1	-1	1	-1	1	-1	
E_1	$\begin{cases} 1 \\ 1 \end{cases}$	$\begin{matrix} \varepsilon \\ \varepsilon^* \end{matrix}$	$\begin{matrix} -\varepsilon^* \\ -\varepsilon \end{matrix}$	$\begin{matrix} -1 \\ -1 \end{matrix}$	$\begin{matrix} -\varepsilon \\ -\varepsilon^* \end{matrix}$	$\begin{matrix} \varepsilon^* \\ \varepsilon \end{matrix}$	
E_2	$\begin{cases} 1 \\ 1 \end{cases}$	$\begin{matrix} -\varepsilon^* \\ -\varepsilon \end{matrix}$	$\begin{matrix} -\varepsilon \\ -\varepsilon^* \end{matrix}$	$\begin{matrix} 1 \\ 1 \end{matrix}$	$\begin{matrix} -\varepsilon^* \\ -\varepsilon \end{matrix}$	$\begin{matrix} -\varepsilon \\ -\varepsilon^* \end{matrix}$	

Table C 2. *Character tables of the dihedral groups D_n ($n = 2,3,4,5,6$)*

$D_2 \equiv V$	E	$C_2(z)$	$C_2(y)$	$C_2(x)$
A	1	1	1	1
B_1	1	1	-1	-1
B_2	1	-1	1	-1
B_3	1	-1	-1	1

D_3	E	$2C_3$	$3C_2$
A_1	1	1	1
A_2	1	1	-1
E	2	-1	0

D_4	E	$2C_4$	$C_4^2 = C_2$	$2C_2'$	$2C_2''$
A_1	1	1	1	1	1
A_2	1	1	1	-1	-1
B_1	1	-1	1	1	-1
B_2	1	-1	1	-1	1
E	2	0	-2	0	0

D_5	E	$2C_5$	$2C_5^2$	$5C_2$
A_1	1	1	1	1
A_2	1	1	1	-1
E_1	2	$2\cos\ 72°$	$2\cos 144°$	0
E_2	2	$2\cos 144°$	$2\cos\ 72°$	0

D_6	E	$2C_6$	$2C_3$	C_2	$3C_2'$	$3C_2''$
A_1	1	1	1	1	1	1
A_2	1	1	1	1	-1	-1
B_1	1	-1	1	-1	1	-1
B_2	1	-1	1	-1	-1	1
E_1	2	1	-1	-2	0	0
E_2	2	-1	-1	2	0	0

Table C 3. *Character tables of the groups* C_{nv} ($n = 2, 3, 4, 5, 6$)

C_{2v}	E	C_2	$\sigma_{v(zx)}$	$\sigma_{v(yz)}$
A_1	1	1	1	1
A_2	1	1	-1	-1
B_1	1	-1	1	-1
B_2	1	-1	-1	1

C_{3v}	E	$2C_3$	$3\sigma_v$
A_1	1	1	1
A_2	1	1	-1
E	2	-1	0

C_{4v}	E	$2C_4$	C_2	$2\sigma_v$	$2\sigma_d$
A_1	1	1	1	1	1
A_2	1	1	1	-1	-1
B_1	1	-1	1	1	-1
B_2	1	-1	1	-1	1
E	2	0	-2	0	0

C_{5v}	E	$2C_5$	$2C_5^2$	$5\sigma_v$
A_1	1	1	1	1
A_2	1	1	1	-1
E_1	2	$2\cos 72°$	$2\cos 144°$	0
E_2	2	$2\cos 144°$	$2\cos 72°$	0

C_{6v}	E	$2C_6$	$2C_3$	C_2	$3\sigma_v$	$3\sigma_d$
A_1	1	1	1	1	1	1
A_2	1	1	1	1	-1	-1
B_1	1	-1	1	-1	1	-1
B_2	1	-1	1	-1	-1	1
E_1	2	1	-1	-2	0	0
E_2	2	-1	-1	2	0	0

Table C 4. *Character tables of the groups* C_{nh} ($n = 2, 3, 4, 5, 6$)

C_{2h}	E	C_2	i	σ_h
A_g	1	1	1	1
B_g	1	-1	1	-1
A_u	1	1	-1	-1
B_u	1	-1	-1	1

C_{3h}	E	C_3	C_3^2	σ_h	S_3	S_3^5	$\varepsilon = e^{2\pi i/3}$
A'	1	1	1	1	1	1	
E'	$\begin{cases}1\\1\end{cases}$	$\begin{matrix}\varepsilon\\\varepsilon^*\end{matrix}$	$\begin{matrix}\varepsilon^*\\\varepsilon\end{matrix}$	$\begin{matrix}1\\1\end{matrix}$	$\begin{matrix}\varepsilon\\\varepsilon^*\end{matrix}$	$\begin{matrix}\varepsilon^*\\\varepsilon\end{matrix}$	
A''	1	1	1	-1	-1	-1	
E''	$\begin{cases}1\\1\end{cases}$	$\begin{matrix}\varepsilon\\\varepsilon^*\end{matrix}$	$\begin{matrix}\varepsilon^*\\\varepsilon\end{matrix}$	$\begin{matrix}-1\\-1\end{matrix}$	$\begin{matrix}-\varepsilon\\-\varepsilon^*\end{matrix}$	$\left.\begin{matrix}-\varepsilon^*\\-\varepsilon\end{matrix}\right\}$	

C_{4h}	E	C_4	C_2	C_4^3	i	S_4^3	σ_h	S_4
A_g	1	1	1	1	1	1	1	1
B_g	1	-1	1	-1	1	-1	1	-1
E_g	$\begin{cases}1\\1\end{cases}$	$\begin{matrix}i\\-i\end{matrix}$	$\begin{matrix}-1\\-1\end{matrix}$	$\begin{matrix}-i\\i\end{matrix}$	$\begin{matrix}1\\1\end{matrix}$	$\begin{matrix}i\\-i\end{matrix}$	$\begin{matrix}-1\\-1\end{matrix}$	$\left.\begin{matrix}-i\\i\end{matrix}\right\}$
A_u	1	1	1	1	-1	-1	-1	-1
B_u	1	-1	1	-1	-1	1	-1	1
E_u	$\begin{cases}1\\1\end{cases}$	$\begin{matrix}i\\-i\end{matrix}$	$\begin{matrix}-1\\-1\end{matrix}$	$\begin{matrix}-i\\i\end{matrix}$	$\begin{matrix}-1\\-1\end{matrix}$	$\begin{matrix}-i\\i\end{matrix}$	$\begin{matrix}1\\1\end{matrix}$	$\left.\begin{matrix}i\\-i\end{matrix}\right\}$

C_{5h}	E	C_5	C_5^2	C_5^3	C_5^4	σ_h	S_5	S_5^7	S_5^3	S_5^9	$\varepsilon = e^{2\pi i/5}$
A'	1	1	1	1	1	1	1	1	1	1	
E_1'	$\begin{cases}1\\1\end{cases}$	$\begin{matrix}\varepsilon\\\varepsilon^*\end{matrix}$	$\begin{matrix}\varepsilon^2\\\varepsilon^{2*}\end{matrix}$	$\begin{matrix}\varepsilon^{2*}\\\varepsilon^2\end{matrix}$	$\begin{matrix}\varepsilon^*\\\varepsilon\end{matrix}$	$\begin{matrix}1\\1\end{matrix}$	$\begin{matrix}\varepsilon\\\varepsilon^*\end{matrix}$	$\begin{matrix}\varepsilon^2\\\varepsilon^{2*}\end{matrix}$	$\begin{matrix}\varepsilon^{2*}\\\varepsilon^2\end{matrix}$	$\left.\begin{matrix}\varepsilon^*\\\varepsilon\end{matrix}\right\}$	
E_2'	$\begin{cases}1\\1\end{cases}$	$\begin{matrix}\varepsilon^2\\\varepsilon^{2*}\end{matrix}$	$\begin{matrix}\varepsilon^*\\\varepsilon\end{matrix}$	$\begin{matrix}\varepsilon\\\varepsilon^*\end{matrix}$	$\begin{matrix}\varepsilon^{2*}\\\varepsilon^2\end{matrix}$	$\begin{matrix}1\\1\end{matrix}$	$\begin{matrix}\varepsilon^2\\\varepsilon^{2*}\end{matrix}$	$\begin{matrix}\varepsilon^*\\\varepsilon\end{matrix}$	$\begin{matrix}\varepsilon\\\varepsilon^*\end{matrix}$	$\left.\begin{matrix}\varepsilon^{2*}\\\varepsilon\end{matrix}\right\}$	
A''	1	1	1	1	1	-1	-1	-1	-1	-1	
E_1''	$\begin{cases}1\\1\end{cases}$	$\begin{matrix}\varepsilon\\\varepsilon^*\end{matrix}$	$\begin{matrix}\varepsilon^2\\\varepsilon^{2*}\end{matrix}$	$\begin{matrix}\varepsilon^{2*}\\\varepsilon^2\end{matrix}$	$\begin{matrix}\varepsilon^*\\\varepsilon\end{matrix}$	$\begin{matrix}-1\\-1\end{matrix}$	$\begin{matrix}-\varepsilon\\-\varepsilon^*\end{matrix}$	$\begin{matrix}-\varepsilon^2\\-\varepsilon^{2*}\end{matrix}$	$\begin{matrix}-\varepsilon^{2*}\\-\varepsilon^2\end{matrix}$	$\left.\begin{matrix}-\varepsilon^*\\-\varepsilon\end{matrix}\right\}$	
E_2''	$\begin{cases}1\\1\end{cases}$	$\begin{matrix}\varepsilon^2\\\varepsilon^{2*}\end{matrix}$	$\begin{matrix}\varepsilon^*\\\varepsilon\end{matrix}$	$\begin{matrix}\varepsilon\\\varepsilon^*\end{matrix}$	$\begin{matrix}\varepsilon^{2*}\\\varepsilon^2\end{matrix}$	$\begin{matrix}-1\\-1\end{matrix}$	$\begin{matrix}-\varepsilon^2\\-\varepsilon^{2*}\end{matrix}$	$\begin{matrix}-\varepsilon^*\\-\varepsilon\end{matrix}$	$\begin{matrix}-\varepsilon\\-\varepsilon^*\end{matrix}$	$\left.\begin{matrix}-\varepsilon^{2*}\\-\varepsilon^2\end{matrix}\right\}$	

Table C 4. (continued)

C_{6h}	E	C_6	C_3	C_2	C_3^5	C_6	i	S_3^5	S_6^5	σ_h	S_6	S_3	$\varepsilon = e^{2\pi i/6}$
A_g	1	1	1	1	1	1	1	1	1	1	1	1	
B_g	1	-1	1	-1	1	-1	1	-1	1	-1	1	-1	
E_{1g} $\begin{cases} \\ \end{cases}$	1	ε	$-\varepsilon^*$	-1	$-\varepsilon$	ε^*	1	ε	$-\varepsilon^*$	-1	$-\varepsilon$	ε^*	
	1	ε^*	$-\varepsilon$	-1	$-\varepsilon^*$	ε	1	ε^*	$-\varepsilon$	-1	$-\varepsilon^*$	ε	
E_{2g} $\begin{cases} \\ \end{cases}$	1	$-\varepsilon^*$	$-\varepsilon$	1	$-\varepsilon^*$	$-\varepsilon$	1	$-\varepsilon^*$	$-\varepsilon$	1	$-\varepsilon^*$	$-\varepsilon$	
	1	$-\varepsilon$	$-\varepsilon^*$	1	$-\varepsilon$	$-\varepsilon^*$	1	$-\varepsilon$	$-\varepsilon^*$	1	$-\varepsilon$	$-\varepsilon^*$	
A_u	1	1	1	1	1	1	-1	-1	-1	-1	-1	-1	
B_u	1	-1	1	-1	1	-1	-1	1	-1	1	-1	1	
E_{1u} $\begin{cases} \\ \end{cases}$	1	ε	$-\varepsilon^*$	-1	$-\varepsilon$	ε^*	-1	$-\varepsilon$	ε^*	1	ε	$-\varepsilon^*$	
	1	ε^*	$-\varepsilon$	-1	$-\varepsilon^*$	ε	-1	$-\varepsilon^*$	ε	1	ε^*	ε	
E_{2u} $\begin{cases} \\ \end{cases}$	1	$-\varepsilon^*$	$-\varepsilon$	1	$-\varepsilon^*$	$-\varepsilon$	-1	ε^*	ε	-1	ε^*	ε	
	1	$-\varepsilon$	$-\varepsilon^*$	1	$-\varepsilon$	$-\varepsilon^*$	-1	ε	ε^*	-1	ε	ε^*	

Table C 5. *Tables of characters of the groups* D_{nh} $(n = 2,3,4,5,6)$

$D_{2h} \equiv V_h$	E	$C_2(z)$	$C_2(y)$	$C_2(x)$	i	$\sigma(xy)$	$\sigma(zx)$	$\sigma(yz)$
A_g	1	1	1	1	1	1	1	1
B_{1g}	1	1	-1	-1	1	1	-1	-1
B_{2g}	1	-1	1	-1	1	-1	1	-1
B_{3g}	1	-1	-1	1	1	-1	-1	1
A_u	1	1	1	1	-1	-1	-1	-1
B_{1u}	1	1	-1	-1	-1	-1	1	1
B_{2u}	1	-1	1	-1	-1	1	-1	1
B_{3u}	1	-1	-1	1	-1	1	1	-1

D_{3h}	E	$2C_3$	$3C_2$	σ_h	$2S_3$	$3\sigma_v$
A_1'	1	1	1	1	1	1
A_2'	1	1	-1	1	1	-1
E'	2	-1	0	2	-1	0
A_1''	1	1	1	-1	-1	-1
A_2''	1	1	-1	-1	-1	1
E''	2	-1	0	-2	1	0

Table C 5. (continued)

D_{4h}	E	$2C_4$	C_2	$2C_2'$	$2C_2''$	i	$2S_4$	σ_h	$2\sigma_v$	$2\sigma_d$
A_{1g}	1	1	1	1	1	1	1	1	1	1
A_{2g}	1	1	1	-1	-1	1	1	1	-1	-1
B_{1g}	1	-1	1	1	-1	1	-1	1	1	-1
B_{2g}	1	-1	1	-1	1	1	-1	1	-1	1
E_g	2	0	-2	0	0	2	0	-2	0	0
A_{1u}	1	1	1	1	1	-1	-1	-1	-1	-1
A_{2u}	1	1	1	-1	-1	-1	-1	-1	1	1
B_{1u}	1	-1	1	1	-1	-1	1	-1	-1	1
B_{2u}	1	-1	1	-1	1	-1	1	-1	1	-1
E_u	2	0	-2	0	0	-2	0	2	0	0

D_{5h}	E	$2C_5$	$2C_5^2$	$5C_2$	σ_h	$2S_5$	$2S_5^3$	$5\sigma_v$
A_1'	1	1	1	1	1	1	1	1
A_2'	1	1	1	-1	1	1	1	-1
E_1'	2	$2\cos 72°$	$2\cos 144°$	0	2	$2\cos 72°$	$2\cos 144°$	0
E_2'	2	$2\cos 144°$	$2\cos 72°$	0	2	$2\cos 144°$	$2\cos 72°$	0
A_1''	1	1	1	1	-1	-1	-1	-1
A_2''	1	1	1	-1	-1	-1	-1	1
E_1''	2	$2\cos 72°$	$2\cos 144°$	0	-2	$-2\cos 72°$	$-2\cos 144°$	0
E_2''	2	$2\cos 144°$	$2\cos 72°$	0	-2	$-2\cos 144°$	$-2\cos 72°$	0

D_{6h}	E	$2C_6$	$2C_3$	C_2	$3C_2'$	$3C_2''$	i	$2S_3$	$2S_6$	σ_h	$3\sigma_d$	$3\sigma_v$
A_{1g}	1	1	1	1	1	1	1	1	1	1	1	1
A_{2g}	1	1	1	1	-1	-1	1	1	1	1	-1	-1
B_{1g}	1	-1	1	-1	1	-1	1	-1	1	-1	1	-1
B_{2g}	1	-1	1	-1	-1	1	1	-1	1	-1	-1	1
E_{1g}	2	1	-1	-2	0	0	2	1	-1	-2	0	0
E_{2g}	2	-1	-1	2	0	0	2	-1	-1	2	0	0
A_{1u}	1	1	1	1	1	1	-1	-1	-1	-1	-1	-1
A_{2u}	1	1	1	1	-1	-1	-1	-1	-1	-1	1	1
B_{1u}	1	-1	1	-1	1	-1	-1	1	-1	1	-1	1
B_{2u}	1	-1	1	-1	-1	1	-1	1	-1	1	1	-1
E_{1u}	2	1	-1	-2	0	0	-2	-1	1	2	0	0
E_{2u}	2	-1	-1	2	0	0	-2	1	1	-2	0	0

Table C 6. *Character tables of the groups D_{nd} ($n = 2, 3, 4, 5, 6$)*

$D_{2d} \equiv V_d$	E	$2S_4$	C_2	$2C_2'$	$2\sigma_d$
A_1	1	1	1	1	1
A_2	1	1	1	-1	-1
B_1	1	-1	1	1	-1
B_2	1	-1	1	-1	1
E	2	0	-2	0	0

Table C 6. (continued)

D_{3_d}	E	$2C_3$	$3C_2$	i	$2S_6$	$3\sigma_d$
A_{1_g}	1	1	1	1	1	1
A_{2_g}	1	1	−1	1	1	−1
E_g	2	−1	0	2	−1	0
A_{1_u}	1	1	1	−1	−1	−1
A_{2_u}	1	1	−1	−1	−1	1
E_u	2	−1	0	−2	1	0

D_{4_d}	E	$2S_8$	$2C_4$	$2S_8^3$	C_2	$4C_2'$	$4\sigma_d$
A_1	1	1	1	1	1	1	1
A_2	1	1	1	1	1	−1	−1
B_1	1	−1	1	−1	1	1	−1
B_2	1	−1	1	−1	1	−1	1
E_1	2	$\sqrt{2}$	0	$-\sqrt{2}$	−2	0	0
E_2	2	0	−2	0	2	0	0
E_3	2	$-\sqrt{2}$	0	$\sqrt{2}$	−2	0	0

D_{5_d}	E	$2C_5^2$	$2C_5$	$5C_2$	i	$2S_{10}^3$	$2S_{10}$	$5\sigma_d$
A_{1_g}	1	1	1	1	1	1	1	1
A_{2_g}	1	1	1	−1	1	1	1	−1
E_{1_g}	2	2 cos 72°	2 cos 144°	0	2	2 cos 72°	2 cos 144°	0
E_{2_g}	2	2 cos 144°	2 cos 72°	0	2	2 cos 144°	2 cos 72°	0
A_{1_u}	1	1	1	1	−1	−1	−1	−1
A_{2_u}	1	1	1	−1	−1	−1	−1	1
E_{1_u}	2	2 cos 72°	2 cos 144°	0	−2	−2 cos 72°	−2 cos 144°	0
E_{2_u}	2	2 cos 144°	2 cos 72°	0	−2	−2 cos 144°	−2 cos 72°	0

D_{6_d}	E	$2S_{12}$	$2C_6$	$2S_4$	$2C_3$	$2S_{12}^5$	C_2	$6C_2'$	$6\sigma_d$
A_1	1	1	1	1	1	1	1	1	1
A_2	1	1	1	1	1	1	1	−1	−1
B_1	1	−1	1	−1	1	−1	1	1	−1
B_2	1	−1	1	−1	1	−1	1	−1	1
E_1	2	$\sqrt{3}$	1	0	−1	$-\sqrt{3}$	−2	0	0
E_2	2	1	−1	−2	−1	1	2	0	0
E_3	2	0	−2	0	2	0	−2	0	0
E_4	2	−1	−1	2	−1	−1	2	0	0
E_5	2	$-\sqrt{3}$	1	0	−1	$\sqrt{3}$	−2	0	0

Table C 7. *Character tables of the groups S_n (n = 4,6,8)*

S_4	E	S_4	C_2	S_4^3
A	1	1	1	1
B	1	-1	1	-1
E	$\begin{cases}1\\1\end{cases}$	$\begin{matrix}i\\-i\end{matrix}$	$\begin{matrix}-1\\-1\end{matrix}$	$\left.\begin{matrix}-i\\i\end{matrix}\right\}$

S_6	E	C_3	C_3^2	i	S_6^5	S_6	$\varepsilon = e^{2\pi i/3}$
A_g	1	1	1	1	1	1	
E_g	$\begin{cases}1\\1\end{cases}$	$\begin{matrix}\varepsilon\\\varepsilon^*\end{matrix}$	$\begin{matrix}\varepsilon^*\\\varepsilon\end{matrix}$	$\begin{matrix}1\\1\end{matrix}$	$\begin{matrix}\varepsilon\\\varepsilon^*\end{matrix}$	$\left.\begin{matrix}\varepsilon^*\\\varepsilon\end{matrix}\right\}$	
A_u	1	1	1	-1	-1	-1	
E_u	$\begin{cases}1\\1\end{cases}$	$\begin{matrix}\varepsilon\\\varepsilon^*\end{matrix}$	$\begin{matrix}\varepsilon^*\\\varepsilon\end{matrix}$	$\begin{matrix}-1\\-1\end{matrix}$	$\begin{matrix}-\varepsilon\\-\varepsilon^*\end{matrix}$	$\left.\begin{matrix}-\varepsilon^*\\-\varepsilon\end{matrix}\right\}$	

S_8	E	S_8	C_4	S_8^3	C_2	S_8^5	C_4	S_8^7	$\varepsilon = e^{2\pi i/8}$
A	1	1	1	1	1	1	1	1	
B	1	-1	1	-1	1	-1	1	-1	
E_1	$\begin{cases}1\\1\end{cases}$	$\begin{matrix}\varepsilon\\\varepsilon^*\end{matrix}$	$\begin{matrix}i\\-i\end{matrix}$	$\begin{matrix}-\varepsilon^*\\-\varepsilon\end{matrix}$	$\begin{matrix}-1\\-1\end{matrix}$	$\begin{matrix}-\varepsilon\\-\varepsilon^*\end{matrix}$	$\begin{matrix}-i\\i\end{matrix}$	$\left.\begin{matrix}\varepsilon^*\\\varepsilon\end{matrix}\right\}$	
E_2	$\begin{cases}1\\1\end{cases}$	$\begin{matrix}i\\-i\end{matrix}$	$\begin{matrix}-1\\-1\end{matrix}$	$\begin{matrix}-i\\i\end{matrix}$	$\begin{matrix}1\\1\end{matrix}$	$\begin{matrix}i\\-i\end{matrix}$	$\begin{matrix}-1\\-1\end{matrix}$	$\left.\begin{matrix}-i\\i\end{matrix}\right\}$	
E_3	$\begin{cases}1\\1\end{cases}$	$\begin{matrix}-\varepsilon^*\\-\varepsilon\end{matrix}$	$\begin{matrix}-i\\i\end{matrix}$	$\begin{matrix}\varepsilon\\\varepsilon^*\end{matrix}$	$\begin{matrix}-1\\-1\end{matrix}$	$\begin{matrix}\varepsilon^*\\\varepsilon\end{matrix}$	$\begin{matrix}i\\-i\end{matrix}$	$\left.\begin{matrix}-\varepsilon\\-\varepsilon^*\end{matrix}\right\}$	

Table C 8. *Character tables of the cubic groups T, T_h, T_d, O, O_h, I, I_h*

T	E	$4C_3$	$4C_3^2$	$3C_2$	$\varepsilon = e^{2\pi i/3}$
A	1	1	1	1	
E	$\begin{cases}1\\1\end{cases}$	$\begin{matrix}\varepsilon\\\varepsilon^*\end{matrix}$	$\begin{matrix}\varepsilon^*\\\varepsilon\end{matrix}$	$\left.\begin{matrix}1\\1\end{matrix}\right\}$	
T	3	0	0	-1	

$T_h = T \times i$; T in T_u.

T_d	E	$8C_3$	$3C_2$	$6S_4$	$6\sigma_d$
O	E	$8C_3$	$3C_2$	$6C_4$	$6C_2'$
A_1	1	1	1	1	1
A_2	1	1	1	-1	-1
E	2	-1	2	0	0
T_1	3	0	-1	1	-1
T_2	3	0	-1	-1	1

$O_h = O \times i$; T in T_{1u}.

Table C 8. (continued)

I	E	$12C_5$	$12C_5^2$	$20C_3$	$15C_2$
A	1	1	1	1	1
T_1	3	$\dfrac{1+\sqrt{5}}{2}$	$\dfrac{1-\sqrt{5}}{2}$	0	-1
T_2	3	$\dfrac{1-\sqrt{5}}{2}$	$\dfrac{1+\sqrt{5}}{2}$	0	-1
G	4	-1	-1	1	0
H	5	0	0	-1	0

$I_h = I \times i$; T in T_{1_u}.

Table C 9. *Character tables of the groups $C_{\infty v}$ and $D_{\infty h}$ for linear molecules*

$C_{\infty v}$	E	$2C_\infty^\varphi$	$\ldots$	$\infty \sigma_v$
A_1	1	1	$\ldots$	1
A_2	1	1	$\ldots$	-1
E_1	2	$2\cos\varphi$	$\ldots$	0
E_2	2	$2\cos 2\varphi$	$\ldots$	0
E_3	2	$2\cos 3\varphi$	$\ldots$	0
$\ldots$	$\ldots$	$\ldots$		$\ldots$

$D^{\infty h}$	E	$2C_\infty^\varphi$	$\ldots$	$\infty \sigma_v$	i	$2S_\infty^\varphi$	$\ldots$	∞C_2
A_{1_g}	1	1	$\ldots$	1	1	1	$\ldots$	1
A_{2_g}	1	1	$\ldots$	-1	1	1	$\ldots$	-1
E_{1_g}	2	$2\cos\varphi$	$\ldots$	0	2	$-2\cos\varphi$	$\ldots$	0
E_{2_g}	2	$2\cos 2\varphi$	$\ldots$	0	2	$2\cos 2\varphi$	$\ldots$	0
$\ldots$	$\ldots$	$\ldots$	$\ldots$	$\ldots$	$\ldots$	$\ldots$	$\ldots$	$\ldots$
A_{1_u}	1	1	$\ldots$	1	-1	-1	$\ldots$	-1
A_{2_u}	1	1	$\ldots$	-1	-1	-1	$\ldots$	1
E_{1_u}	2	$2\cos\varphi$	$\ldots$	0	-2	$2\cos\varphi$	$\ldots$	0
E_{2_u}	2	$2\cos 2\varphi$	$\ldots$	0	-2	$-2\cos 2\varphi$	$\ldots$	0
$\ldots$	$\ldots$	$\ldots$	$\ldots$	$\ldots$	$\ldots$	$\ldots$	$\ldots$	$\ldots$

References

1. Woodman, C. M.: Mol. Phys. **11**, 109 (1966).
2. Emsley, J. W., J. Feeney, and L. H. Sutcliffe: High Resolution NMR Spectroscopy. London-New York-Paris: Pergamon Press 1965.
3. Cotton, F. A.: Chemical Applications of Group Theory. New York and London: Interscience Publishers, John Wiley & Sons, Inc. 1964.
4. Longuet-Higgins, H. C.: Mol. Phys. **6**, 445 (1963).
5. Hougen, J. T.: J. Chem. Phys. **39**, 358 (1963); – I.U.P.A.C. VIIIth European Congress on Molecular Spectroscopy; p. 481. London: Butterworths 1965.
6. Altmann, S. L.: Proc. Roy. Soc. **Ser A 298**, 184 (1967).
7. Lynden-Bell, R. M.: Unpublished research.
8. Woodman, C. M.: Mol. Phys. **13**, 365 (1967).
9. Gutowsky, H. S., D. W. McCall, and C. P. Slichter: J. Chem. Phys. **21**, 279 (1953).
10. Saupe, A., and J. Nehring: J. Chem. Phys. **47**, 5459 (1967).
11. Harris, R. K., and C. M. Woodman: Mol. Phys. **10**, 437 (1966).
12. Pople, J. A., W. G. Schneider, and H. J. Bernstein: High Resolution NMR. New York-Toronto-London: McGraw-Hill Book Co., Inc. 1959.
13. McConnell, H. M., A. D. McLean, and C. A. Reilly: J. Chem. Phys. **23**, 1152 (1955).
14. Jones, R. G., R. C. Hirst, and H. J. Bernstein: Can. J. Chem. **43**, 683 (1965).
15. –, and S. M. Walker: Mol. Phys. **10**, 349 (1966).
16. –, and P. Partington: Unpublished research.
17. Acrivos, J. V.: Mol. Phys. **5**, 1 (1962).
18. Wilkinson, G. and T. S. Piper: J. Inorg. Nucl. Chem. **2**, 23 (1956). Piper, T. S., and G. Wilkinson: ibid., **3**, 104 (1956).
19. Pople, J. A., W. G. Schneider, and H. J. Bernstein: Can. J. Chem. **35**, 1060 (1957).
20. Diehl, P., R. G. Jones, and H. J. Bernstein: Can. J. Chem. **43**, 81 (1965).
21. –, C. L. Khetrapal, and H. P. Kellerhals: Mol. Phys. **15**, 3337 (1968).
22. Lawrenson, I. J.: J. Chem. Soc. 1117 (1965).
23. –, and R. G. Jones: J. Chem. Soc. 3336 (1967).
24. Bulthuis, J., J. Gerritsen, C. W. Hilbers, and C. Maclean: Recueil des Pays Bas. **87**, 417 (1968).
25. Haig, C. W.: Private communication.
26. Diehl, P., R. K. Harris, and R. G. Jones: 'Sub-spectral Analysis', Progress in NMR Spectroscopy **3**, 1 (1967).
27. – Helv. Chim. Acta **48**, 567 (1965).
28. Aitken, A. C.: Determinants and Matrices. New York and London: John Wiley & Sons, Inc. 1958. – R. P. Baumann: Absorption Spectroscopy, Appendix I. New York and London: John Wiley & Sons, Inc. 1963.
29. Bright-Wilson, E., Jr.: J. Chem. Phys. **27**, 60 (1957).
30. Grimley, T. B.: Mol. Phys. **6**, 329 (1963).
31. Lynden-Bell, R. M.: Mol. Phys. **15**, 523 (1968).
32. Reddy, G. S., and R. Schmutzler: Inorg. Chem. **6**, 823 (1967).
33. Nixon, J. F.: J. Chem. Soc. **A**, 1136 (1967).
34. Woodman, C. M.: Private communication.
35. Wilson, E. B., J. C. Decius, and P. D. Cross: Molecular Vibrations. New York: McGraw-Hill Book Company Inc. 1955.
36. Sandorfy, C.: Electronic Spectra and Quantum Chemistry. New Jersey: Prentice-Hall Inc. 1964.